INEQUALITY

INE**Q**UALITY

Darwinian Evolution
and
Disparity in the Wealth of Nations

Harold Lewis Longaker

Napoleon Avenue

Published by Napoleon Avenue
5500 Prytania St., #114
New Orleans, LA 70115

www.napoleonavenuepublishing.com

Printed in the United States of America

20 19 18 17 16   1 2 3 4 5

ISBN: 978-0-9979617-0-6 (pbk)

Library of Congress Control Number: 2016916762

To those who have been an important part of my life but
have passed ahead, especially Rebecca

Last, but always first, for Lynn

We are self-made, independent, alone and fragile. Self-understanding is what counts for long-term survival, both for individuals and for the species.

—Edward O. Wilson, *The Riddle of the Human Species*

Contents

Figures

Tables

Preface

The origins of this book are in Gregory Clark's *A Farewell to Alms: A Brief Economic History of the World.* Clark, the economic historian that he is, asks the question, why did the Industrial Revolution start on that cold, damp island clinging to the far edge of the Eurasian landmass? Why was it England and not China, Germany, or Argentina?

If the number of reviews is a measure of a book's importance, Clark's, with at least seventy-nine in diverse media such as the *New York Times, Forbes, Science, Harvard Business Review,* and *Cato Journal,* is important. For our purposes, however, it is important because Clark implicitly uses Darwinian Theory to explain a change in a human condition, that of populations gaining in economic prosperity. He reports that rich English testators of the preindustrial period left behind more progeny than poor ones did, making them, in a biological, Darwinian sense, fitter.

In the preindustrial period, population growth was essentially zero. This translates to the scions of the rich having to fall down the demographic hill to fill in the spaces left by the less reproductively fit poor, a conclusion required by simple mathematics. The scions of the rich who fill in the lower brackets of the hierarchy bring with them their heritable traits, sourced from their rich fathers. The important consequence is that the genes and heritable cultural traits leading to individuals becoming rich increase in the population, making the population better at wealth creation. This is Darwinian evolution at work.

With this important observation, Clark develops an answer to his question, but unfortunately, it ends up incomplete. The question is wider and deeper than his approach. Clark is an economic historian, and perhaps he lacks sufficient breadth of expertise to attack the question fully. His question requires going further back in time to an era before the conquest of Britain by the Roman Empire. This is the temporal distance required to understand how the rich English testator came to be in the first place, an effort requiring anthropology.

Adding to this incompleteness is a lack of cultural diversity, as Clark's focus is primarily on the English. Achieving a proper answer to the type of question he asks begs for a comparative, relative approach. Cross-cultural comparisons enable detecting cause and effect. Using this tool, we can beat one against the other, detect differences, and create understanding. Furthermore, differences and their causes are far easier to discern, much more so than on an absolute, nonrelative basis. For a comparative analysis, Clark would have to bring in other nations and their cultures, and this, too, requires anthropology.

Inequality: Darwinian Evolution and Disparity in the Wealth of Nations addresses a derivative question: why was it that only some nations grabbed the brass ring of industrialization that leads to prosperity? The fruit of the Industrial Revolution was available to all takers, yet some did not partake, and they remain, even after all these intervening years, as poor now as then. The reason for this book, and its goal, is to answer the question, why are some nations rich, while others remain poor?

Similar to Clark's, this book also needs the two dimensions of depth of time and width of cultural diversity—and a third as well: the vista of height provided by a multiplicity of disciplines. We have mentioned two, economic history and anthropology, but our topic is wider than these disciplines can encompass. Employing only one or the other would unknowingly leave strings untied and obscure corners unexplored. For our story to be complete, we will need psychometrics, because part of the story considers intelligence as measured by IQ. We will need some psychiatry, as we will discuss McClelland's three-part Need Theory.

Gene–culture coevolution or Dual Inheritance Theory is also part of the explanation, and for this, we need the input of someone like Pete Richerson, an expert in social cultural evolution and human ecology. Some of the story concerns our being a social animal, a domain of sociologists, and because our topic of rich and poor has more than a tint of economics, we need that expertise as well. We could make the list longer, but we have made our point. We need the depth of time, the width of multiple cultures, and the presence of several disciplines to get us where we want to go.

The last of the preceding points suggests a requirement for a more integrated approach, and this leads us to the second important seed of this book, Edward O. Wilson's *Consilience: The Unity of Knowledge*. *Consilience* is a neatly precise word, but I expect some have no idea what it means. We can deduce from the title of Wilson's book that it probably means "unity of knowledge." It does, but there is more:

> Consilience, literally a "jumping together" of knowledge by the linking of facts and fact-based theory across disciplines to create a common groundwork of explanation.[1]

Developing an answer to our question requires linking together perspectives, as well as knowledge, from various learned groups and their associated paradigms. We need to stitch together bits and pieces of fact and theory from a variety of disciplines to form the coherent, whole cloth.

Academics, almost by definition, are specialists, not generalists. They understand their respective disciplines to great depth but may lack a wide vision. This is not a criticism; given what they do, it must be so. However, considering the number of disciplines needed for our question, we need a generalist, one trading off depth for width. We can define the requirement this way: "Jack of all trades, master of none, though oftentimes better than master of one."[2]

I might not be the best person to write this book, but I am close. I have had an assortment of careers. For part, I was a biological oceanographer, during which time I started to develop a lifelong interest in behavior and its reasons. For the rest, I worked in a variety of technical arenas, developing expertise in nuclear power engineering, geodesy, project management of offshore operations, GPS systems engineering, data and wireless communication, exploration geophysics, and market development. Over the years, I put my name as inventor on eighteen U.S. patents. I have looked at the world from a multitude of perspectives and have learned to appreciate and utilize a consilient approach to complex problems.

In addition to the lack of consilience in academic professions, there is another reason that I, rather than someone in academia, wrote this book. The question requires inquiring why populations achieve different economic outcomes, and this necessitates examining differences between populations. Academia does not allow such inquiry, because some consider even asking the question to be racist. The current environment in academia absolutely prevents any inquiry having even the possibility of a racist tint; anyone so trying will have her career terminated. Furthermore, why take that risk while there are so many other interesting questions awaiting inquiry?

There is a second case for a consilient approach, or an approach with a wide view. Getting to our answer requires inquiries into the nature of early societies. This is the domain of anthropologists, and unfortunately, owing to their "own confusion regarding what type of science anthropology really was," they have made an "epistemological misstep" leading to an "under-use of . . . evolutionary theory."[3] Likely because of the aforementioned, they seem to have missed an important aspect of early societies. What multiple ethnographic studies showed, but failed to comprehend, was that the sociality of early human societies bears strong resemblance to that of the social insects—the bees, ants, and termites.

This failure results in missing a critical aspect of early human societies. Supersociality, the superorganism, and group selection are all of a piece, and that is where the evidence points. If correct, this nature represents a significant difference from currently held views. It suggests that individuals labored for the good of the group, not selves, while in turn, the purpose of group behavior was the individual's survival; we were superorganismic, and group natural selection was evolution's *modus operandi*. By exposing the path never taken, the foregoing leads to a novel foundation on which to build a Darwinian model, and that path is where this book goes.

The starting intent of this book's project was to address the rich–poor dichotomy with the theoretical foundation of Darwinian Theory and to take the known and translate it into a coherent picture. Obviously, this changes with the new path.

Frankly, when I got to this part of my research, I was surprised, as I did not expect to find such discoveries. Their consequence, by presenting a novel view of the social nature of our ancestors, makes this book more than a study of the poor and the rich. It becomes a story of our development, starting at a hitherto unexplored point and progressing to where we are today. Importantly, traversing this unexplored path lends clarity to our understanding of us.

This unexpected, new view should be in an academic journal, and perhaps it will be. However, by the time I realized I was pursuing a path with ideas not yet "sanctioned" by peer review, this book was well under way. My choices were either to stop, get the new ideas into the literature, and then restart, or to continue. Obviously, I continued. My rationale is that most of the conclusions developed in this book do not require these nonreviewed ideas. They could stand without them, but not as clearly. Including them makes the analysis more understandable.

There is a second point missed by anthropologists, one related to missing the first. It is the aspect of the Urban Revolution during which we transition from rural to urban life. In this transition, we lose our supersocial nature, that of being similar to social insects, and become more like modern man, not quite as social but still retaining several traits of our prior supersocial existence. This also markedly affects the development of our answer.

In the process of developing an explanation for "why do some remain poor," the new ideas discussed in the preceding paragraphs came into being. Explaining the rich also develops new ideas, but in this case, there is a difference. We get to the answer of "why the rich" by starting at Clark's observation and following Darwinian Theory to its conclusion. This is new and interesting, but not a "discovery." However, what falls out in the process of model development is a discovery that turns out to be another nonreviewed idea: a Darwinian explanation for why some populations have higher IQs than others do. The conclusion is that the same process leading to economic prosperity also leads to increased intelligence; populations become wealthier and smarter.

The new ideas discussed here lend a sense of newness to this book, but they are not its only or even primary source. From one perspective, this book is not new because, excluding the noted "discoveries," it primarily stitches together accepted concepts. It is new because of its consilient approach, with its uniqueness coming from applying Darwinian Theory of one paradigm to anthropological observations of another. Anthropologists tell us what happened, but not necessarily the whys—those come from Darwin. This approach in and of itself generates several new ideas.

To some this book will appear sexist, and to a degree, it probably is. It is not possible to write about the history of humans and their development without writing about things sexist. That is the way the world was and, in many ways, still is. That different genders have different reproductive strategies should not translate into one gender being "sexist." The intent of the so-called sexist writing that exists herein, if any, is to be consistent with the times and places being discussed. If it appears otherwise, I did not do my job.

Many nonfiction books are constructed top down, working their way down by decomposing the complex into smaller and more understandable pieces. This book is different, using a bottom-up synthesis to create a higher-order, complex structure by stitching together basic building blocks from multiple disciplines.

As discussed, this book contains an element of discovery, and the order of the narrative's presentation follows discovery's path. I did this with the expectation that the reader will better understand the reasoning involved, even though some might just want the answer and consider such an approach confounding.

Somewhere toward the end of developing my model and this book, a persistent question started to arise. Why am I writing this thing? It had become obvious that what I was doing, others should have done ten, even twenty years ago. Writing this preface at the end of my labor helped me answer that question. The reasons are several, all discussed earlier. However, the one that stands out the most is consilience. I found it to be an extraordinarily powerful tool, yet one grossly underutilized in genres of this type.

The consequence of writing a consilient book is that it cannot use the arcane language of a single learned discipline; too many specialties are involved. Additionally, complicated concepts should be broad and unembellished, and an intended audience of everyman requires common, everyday verbiage. The specialist might prefer a more nuanced approach, employing more details. However, nuances can get in the way of a novice's understanding, and details can obfuscate. Getting the main point is the point.

At the end of the day, this book is mine, and I own all the errors. After they have been resolved, I expect to find remaining real differences in opinion. That is where the interesting stuff begins, as differences in opinion are, in my view, the precursors to understanding.

Acknowledgments

Finishing the writing of a book is a beginning. It is at that time when fresh eyes get the opportunity to examine and detect incorrect grammar, punctuation, and word usage and, for the first time, others the chance to examine and challenge heretofore unchallenged ideas. This is the juncture where an author needs all the help he can find. I am very thankful it was there.

I tend to write as I talk, without too much concern for tone. I say what I think, sometimes without thinking. I thank both my wife, Lynn, and my good friend Les Riess for helping guide to moderation the sometimes intemperate nature of my writing. Along the same vein, writing a potentially controversial book with the requisite sense of diplomacy is not natural for a tone-deaf person like myself. I am grateful to Kim Benson for her thoughtful review, where she marked language on the wrong side of political correctness. However, some of what she found remains, because sometimes, political incorrectness is a useful tool in making a point.

This book is concerned with civilization in both the anthropological and political science senses. When these disciplines are mixed and "civilization" is the topic, confusion abounds. Bruce Peetz was very helpful in identifying this problem and assisting in unwinding the associated complexity. Bruce often acted as a sounding board, and when he did, he always knew what was right. I cannot thank him enough for his time and especially his insight.

Wealth and income inequality are standard fare for economic discussions. *Inequality* shifts them from their usual habitat to evolutionary biology, giving them added dimensions that had to be carefully parsed. David Bird identified this issue and aided greatly in this esoteric task. He helped make a major point of the book far more understandable. David: thank you.

English is a rule-based discipline in which I admittedly lack proper knowledge. My copy editor, Holly Monteith, knows the rules and, importantly, how to apply them without changing voice. She, too, deserves hearty thanks.

My partner and wife made by far the greatest contribution to this endeavor. For the six years it took to develop and write *Inequality,* she had to put up with me sitting at my desk in a corner of her kitchen. She was the one who, at breakfast, listened to the epiphanies I had "discovered" at 3:00 A.M. the night before. Without Lynn, this book would never have come to fruition. Thank you for this, for all the wonderful years, and for so much more.

CHAPTER 1

Introduction

◇◇◇◇◇◇◇◇◇◇◇◇◇◇◇◇◇◇◇◇◇◇◇◇◇◇◇◇◇◇◇◇

Tolerance and the Uncertainty of Truths

For everything, there is a reason, and for this book, it is the question of why some nations are rich while others remain poor. With varying perspectives, this has been the topic of others, but none has approached this question with a sociobiological or Darwinian paradigm. This is what we are uniquely going to do; it is the reason for our journey.

This book inquires into matters that some might find offensive, as it is concerned with people and their differences, cultural and genetic. I am writing this in a time of cultural conflicts, pitting the ideology of progressivism against that of conservatism. Each holds as absolute its ideas and beliefs; their mutual intransience hinders political and social progress.

To preclude these conflicts from spilling into this book, and clouding its account, I proffer a mitigating perspective. It is from the mathematician Jacob Bronowski and his seminal documentary *The Ascent of Man*. In it he proposes a tolerance of ideas, but not because we ought to be tolerant. It is the uncertainty of our respective countervailing truths that demands this tolerance. Ideas and beliefs, along with their co-conspirators, "truths" and "facts," contain uncertainty, even if it is at some relatively insignificant level:

The Principle of Uncertainty is a bad name. In science, or outside of it, we are not uncertain; our knowledge is merely confined, within a certain tolerance. We should call it the Principle of Tolerance. And I propose that name in two senses. First, in the engineering sense: Science has progressed, step by step, the most successful enterprise in the ascent of man, because it has understood that the exchange of information between man and nature, and man and man, can only take place with a certain tolerance. But second, I also use the word, passionately, about the real world. All knowledge—all information between human beings—can only be exchanged within a play of tolerance. And that is true whether the exchange is in science, or in literature, or in religion, or in politics, or in any form of thought that aspires to dogma. It's a major tragedy of my lifetime and yours that scientists were refining, to the most exquisite precision, the Principle of Tolerance—and turning their backs on the fact that all around them, tolerance was crashing to the ground beyond repair. The Principle of Uncertainty or, in my phrase, the Principle of Tolerance, fixed once and for all the realization that all knowledge is limited. It is an irony of history that at the very time when this was being worked out, there should rise, under Hitler in Germany and other tyrants elsewhere, a counter-conception: a principle of monstrous certainty. When the future looks back on the 1930s, it will think of them as a crucial confrontation of culture as I have been expounding it—the ascent of man against the throwback to the despots' belief that they have absolute certainty.[1]

This book represents truth as I understand it, and any offense found is unintended. In my world, truth cannot offend. It can be awkward, uncomfortable, unwanted, or inconvenient. In addition, if only apparent, it can be wrong, where this possibility of wrongness extends to my understanding. If you hold other truths, I offer no apology and, in return, expect none.

In our narrative, we are going to examine both the why and the how of differences between populations. Unfortunately, some consider as racist any discussion of between-group differences. Furthermore, we are going to use the science of Darwinian Theory as our primary tool, and in doing so, I recognize that some will refer to this tome as "scientific racism." Those so considering would be wrongheaded, but if the book is so perceived, so be it.

There is also a political nature to the account; it ignores and denies some of the gospels of both the progressive left and the religious right. Many of the religious right do not believe in evolution, and a cornerstone of this book is that humans are an evolved species, just like all the rest of the animals. We have no special rights or privileges other than what we have obtained through evolution, and these are considerable.

On the left, a tenet of many liberals is the Marxian one that all humans are equal and, given identical environments, all can and should achieve equal outcomes. A thesis of this book is that humans are not equal and, even in equal environments, will not achieve equal outcomes. Inequalities exist not only between individuals but also between populations.

A second, more important thesis is that differences have reasons; unveiling them is a purpose of this book. Much is dedicated to understanding the why of differences; the how falls out in the process. For our purposes, processes leading to differentiation are more important than the difference itself. With knowledge of process, we understand as well as gain the ability to predict.

Not only is our subject controversial; so is the tool we will use— sociobiology, or human behavioral ecology. This discipline came into being with Edward O. Wilson's 1975 book *Sociobiology: The New Synthesis* and is the systematic study of the biological basis of social behavior. It is the foundation of this book and brings to our study tools we can use to understand processes leading to differences.

Some of the "science of humans" has more than a mere tint of politics coloring what should be an uncolorable set of scientific disciplines. This includes sociobiology, where the firestorm created by its publication evidences the source of a particular ideological taint. Wilson best expresses its ideological nature in the following quotation:[2]

> Who were the critics, and why were they so offended? Their rank includes the last of the Marxist intellectuals, most prominently represented by Stephen J. Gould and Richard C. Lewontin. They disliked the idea, to put it mildly, that human nature could have any genetic basis at all. They championed the opposing view that the developing human brain is a tabula rasa. The only human nature, they said, is an indefinitely flexible mind. Theirs was the standard political position

taken by Marxists from the late 1920s forward: the ideal political economy is socialism, and the tabula rasa mind of people can be fitted to it. A mind arising from a genetic nature might not prove conformable. Since socialism is the supreme good to be sought, tabula rasa it must be. As Lewontin, Steven Rose, and Leon J. Kaman frankly expressed the matter in *Not in Our Genes* (1984): "We share a commitment to the prospect of the creation of a more socially just—a socialist—society. And we recognize that a critical science is an integral part of the struggle to create the society, just as we also believe that the social function of much of today's science is to hinder the creation of that society by acting to preserve in interests of the dominant class, gender, and race."

That was in 1984—an apostate Orwellian date. The argument for a political test of scientific knowledge lost its strength with the collapse of world socialism and the end of the Cold War. To my knowledge it has not been heard since.[3]

In the 1970s, when the human sociobiology controversy still waxed hot, however, the Old Marxists were joined and greatly strengthened by members of the new left in a second objection, this time centered on social justice. If genes prescribe human nature, they said, then it follows that ineradicable differences in personality and ability also might exist. Such a possibility cannot be tolerated. At least, its discussion cannot be tolerated, said the critics, because it tilts thinking onto a slippery slope down which mankind easily descends to racism, sexism, class oppression, colonialism, and perhaps—worst of all—capitalism! As the century closes, this dispute has been settled. Genetically based variation in individual personality and intelligence has been conclusively demonstrated, although statistical racial differences, if any, remain unproven. At the same time, all the projected evils except capitalism have begun to diminish worldwide. None of the change can be ascribed to human behavioral genetics or sociobiology. Capitalism may yet fall—who can predict history?—But, given the overwhelming evidence at hand, the hereditary framework of human nature seems permanently secure.[4]

Sociobiology is the progenitor of new scientific disciplines such as evolutionary developmental psychology, human behavioral ecology, and evolutionary anthropology. Given that these sprang from Wilson's sociobiology, it should be obvious that he and his supporters won.

Wilson's victory, along with the resulting new disciplines, suggests that the classical social sciences of psychology, anthropology, and sociology have been lacking. The new, Darwinian disciplines of evolutionary psychology, evolutionary anthropology, and evolutionary sociology are the evidence. Assuming a free market, the marketplace of ideas will determine winners and losers. However, in the politically correct atmosphere of modern academia, this assumption is in serious doubt.

Preview of Our Journey's Vista

Before we start, we want to look at two other similar journeys, but taken with different perspectives. We will find that they provide context and reference when we take our journey. The first is the perspective of the historian, for which we will depend on *The Wealth and Poverty of Nations: Why Some Are So Rich and Some So Poor* by the historian David S. Landes. The perspective of the second is geography or environmental determinism, and for this, our reference will be the Pulitzer Prize–winning *Guns, Germs, and Steel: The Fates of Human Societies* by Jared Diamond.

The historian's view. Landes's book is greatly enjoyable. He combines a wonderful breadth of knowledge with superb wordsmithship to create a very pleasurable read. It is a broad, nuanced study of the history of world economics from the start of the Age of Discovery at the end of the fifteenth century onward.

Some have faulted it for being Eurocentric. It is, but it has another fault, one worth addressing. Actually, it is a general fault of other books on this topic. Landes spends most of his effort on the economic winners, trying to understand why they won. He spends little time on the losers.

If losing were only an absence of winning, discussing losers would not be required. However, we will learn on our journey that losing is not necessarily the case of not winning but more like never playing the game. Understanding requires addressing the problem in its entirety. We need to understand that being poor is not always the absence of being rich; it is more complicated.

Historians are like teachers and programmers. Teachers, those educated in the art and science of teaching, know how to teach. They do not know what to teach; for that, they need additional education. Computer programmers know how to write code and, similar to teachers, lack knowledge of what to code. Technology companies do not hire programmers; they hire scientists and engineers, who, by the way, know how to code. Historians know the paradigm of "history" but typically lack expertise in economics, psychology, or other applicable disciplines required to fully explain the why of history. They can only write to the how.

There is no intent to be disparaging of those disciplines, only to show limitations. We all have our role, and theirs has equal importance to others. The point being made is that we should not expect a complete or in-depth understanding from a historian, nor, for that matter, from any single discipline. Achieving a full understanding of complex matters, especially those having a human component, often requires a multidisciplinary, consilient approach.

Given its approach, Landes's book cannot achieve the depth of understanding required by the problem. Nevertheless, it can and does bring out some important top-level issues. Landes's concluding thesis is that differences in economic outcomes are due to differences between cultures:

> If we learn anything from the history of economic development, it is that culture makes all the difference. (Here Max Weber was right on.) Witness the enterprise of expatriate minorities—the Chinese in East and Southeast Asia, Indians in East Africa, Lebanese in West Africa, Jews, and Calvinists throughout much of Europe, and on and on. Yet culture, in the sense of inner values and attitudes that guide a population, frightens scholars. It has a sulfuric odor of race and inheritance, an air of immutability. In thoughtful moments, economists and social scientists recognize that this is not true, and indeed salute examples of cultural change for the better while deploring changes for the worse. But applauding or deploring implies the passivity of the viewer—an inability to use knowledge to shape people and things. The technicians would rather do: change interest rates, free up trade, alter political institutions, manage. Besides, criticisms of culture cut close to the ego, injure identity and self-esteem. Coming from outsiders, such animadversions, however tactful and indirect, stink of condescension. Benevolent improvers have learned to steer clear.[5]

The preceding paragraph covers all the bases:

1. The answer to the question "why the difference in wealth between nations?" lies with the differences in cultures of the populations.

2. Examples are in the economic successes of various dominant minorities.

3. This answer is an anathema to some, as it contains the determinism of genetically based answers. It is also racist; not that it is, but rather those with strong disagreements will call it so.

4. The cognoscenti can appreciate even an anathema, provided it has a foundation of truth.

5. Many are satisfied with nonmeaningful economic adjustments while avoiding the underlying, unpleasant truth. "Forgive the debt and let's repeat whatever it was we did all over again. At least we are doing something!"

6. An economist from a "superior" rich nation criticizing the culture of an "inferior" poor country can expect a black eye, both literally and figuratively.

Landes's presentation is not monothematic; it is about more than just culture. He includes as players technology, institutions, and geography. However, in the main, his conclusion is that "it is culture," but it is not simply culture. If we boil down his conclusion, we find a *glace viande* made with the sweat of hard work:

We want things to be sweet; too many of us work to live and live to be happy. Nothing wrong with that; it just does not promote high productivity. You want high productivity? Then you should live to work and get happiness as a by-product.[6]

Not easy. The people who live to work are a small and fortunate elite. But it is an elite open to newcomers, self-selected, the kind of people who accentuate the positive.[7]

The one lesson that emerges is the need to keep trying. No miracles. No perfection. No millennium. No apocalypse. We must cultivate a

> skeptical faith, avoid dogma, listen and watch well, try to clarify and define ends, the better to choose means.[8]

"Hard work" is too simple of an answer. A nuanced examination will show that there is more to "hard work" than just hard work. We can examine two economically successful groups, Protestant and Confucian, to see why.

The Confucian nations of Hong Kong, Taiwan, South Korea, and Singapore are all toward the top of the GDP list, and the remaining two, China and Vietnam, are rapidly scaling the economic ladder. "Work ethic" is not part of the Confucian ethos. For example, it used to be a sign of status for a Chinese man to have a very long pinky fingernail, as it showed that he did no manual labor. Confucians shun hard work and expend minimal effort to extract maximum gain. The Confucian culture works smarter, not harder; they take the less arduous way to the top. Why write software when all it takes is a simple COPY command? Hard work of itself is not good. There is another explanation for their success.

Max Weber, in his 1904 book *The Protestant Ethic and the Spirit of Capitalism,* ascribed the development of capitalism in England to the work ethic of the Calvinists. According to Landes, the Calvinists took their belief in predestination and "eventually converted it into a secular code of behavior: hard work, honesty, seriousness, the thrifty use of money and time."[9] For Calvinists, work for the sake of work was the defining drive. Calvinist laborers, merchants, and lawyers all worked hard, independently of any reward aside from hard work itself.

Does the foregoing mean that the economic success of many Confucian nations has nothing to do with hard work; is it just working smarter, not harder? We can find the answer to this dichotomy in Lucian Pye's "Asian Values."[10]

According to Pye, the "need for achievement" (N-Ach) is an important Chinese cultural value. It comes from the Confucian ethic of self-improvement, manifested in the teaching to Chinese children of the importance of striving for success. Because of this cultural learning process, Chinese children also feel shame when they do not meet their parents' expectations.

The N-Ach is a subset of the three-part Need Theory[11] developed by psychologist David McClelland. The other two needs are affiliation (N-Affil) and power (N-Pow). Individuals with a high level of N-Ach have, among other propensities, a strong desire for achieving significant accomplishments. If hard work is required for achieving, then hard work it is.

Calvinists work hard for the sake of work, and Chinese work hard for significant accomplishments. However, the Chinese will take the easiest path to obtain the accomplishment, whereas the "easy path" is not on the Calvinist's list of virtuous activities.

Landes's concluding recommendation:

> And what of the poor themselves? History tells us that the most successful cures for poverty come from within. Foreign aid can help, but like windfall wealth, can also hurt. It can discourage effort and plant a crippling sense of incapacity. As the African saying has it, "The hand that receives is always under the one that gives." No, what counts is work, thrift, honesty, patience, tenacity. To people haunted by misery and hunger, that may add up to selfish indifference. But at bottom, no empowerment is so effective as self-empowerment.[12]

The answer is simple—too much so—and for this the historian's answer is found wanting. It is at too high a level to have practical value; we must get deeper into the bowels of the beast to find our answers. However, its basic truth is undeniable. The culture of hard work leads to economic success. People who live to work will economically out-perform those who work to live.

Geography's perspective. As stated earlier, the purpose in developing this book was to address the economic disparity found between nations, and this it does. However, it also winds up with the unintended consequence of being a counterpoint to *Guns, Germs, and Steel*. The perspective of Diamond's book is geography or environmental determinism. His central thesis is that the differences we observe in economic outcomes between people are due to differences in their geographies, never to any differences between people. The underlying thesis of this book is that different economic outcomes result from people being different, a point precisely counter to Diamond's. Is it people or their environment? This is point and counterpoint.

If you were to ask Diamond why the West is so economically advanced, he could well reply, because of the luck of the draw (of their favorable geography). The people of Western Europe and their offspring in America, Canada, New Zealand, and Australia were just plain lucky for being born in an environment rich in resources; that is all there is to it, period! There is nothing superior here.

Diamond succinctly states his thesis in the following quotation: "History followed different courses for different peoples because of differences among peoples' environments, not because of biological differences among the peoples themselves."[13]

The geography or environmental determinism thesis states that the economic success of the West is due to availability of resources. Societies like those of sub-Saharan Africa that have not fared well lacked a comparable amount. Diamond's thesis is that resource inequality between human populations over the long course of their development has led to economic inequality.

An example of this inequality is the distribution of large (weighing more than one hundred pounds) domesticable animals, an assemblage including cattle, pigs, and horses. For this class, Eurasia takes the brass ring. It has seventy-two candidate species, and of those, thirteen have been domesticated. Sub-Saharan Africa has fifty-one, and none has been domesticated. The corresponding numbers for the Americas are twenty-four species with only one, the llama, achieving domestication. At the bottom of the list is Australia, having only a single candidate, the kangaroo.[14]

A big animal can provide lots of protein, but so can a large brood of chickens. However, some big animals, such as sheep and cattle, do provide the means of converting nominally unusable grass energy to usable protein. Big animals provide transportation. Only the Eurasians were able to ride to work or war; the rest of the world had to walk. In addition, big animals give leverage to human innovation—the plow is such an example.

There is truth in Diamond's thesis, just not exactly for the reason he espouses. Additionally, his story has three flaws. We will discuss them all, truths and flaws. Reasons for differences in economic outcomes between people are at the core of his book. His succinct summary both presents his thesis and advertises its flaws. He assigns to geography the role of principal actor in determining human outcomes.

Economic activity is human, and that, not geography, must be the proximate determinant of economic outcomes. Geography's role is constrained to shaper and molder. People make the money, and geography has a role in making the people. Geography has its importance, but not in the manner or to the extent Diamond advocates.

Geography's circumscribed role is to set the evolutionary environment of people and their populations. Evolutionary theory predicts that different geographies presenting different evolutionary environments should result in different evolutionary outcomes. They have, and these different evolutionary outcomes have resulted in differences between populations, with these differing populations in turn producing different outcomes, economic and otherwise. This is where the strength of environmental determinism really lies.

Diamond's journey. Diamond develops his thesis of environmental determinism starting at the end of the last ice age, approximately thirteen thousand years ago. By that time, we had migrated from our homeland of Africa to most of the major islands and to all the continents, except Antarctica. We made our living as hunter-gathers and were organized in small kinship groups or clans consisting of a few tens of individuals. These groups were egalitarian, without any chief or big man; group consensus ruled.

To arrive at our current condition of personal injury lawyers, evolutionary psychologists, web designers, and media celebrities, we had to take a major, revolutionary first step, the Neolithic Revolution. We settled down as agriculturists. Sure, some of us remained as hunter-gatherers, and some went in for ranching, but being dirt farmers was the predominant path. Dirt farming ties people to the land, an important shackle that forces the change from nomadic foraging to stationary farming. We settled down, literally.

We also started to lose our egalitarian political ways. In self-defense, bigger is better, and to increase security, clans conjoined to form larger tribes. They acquired chiefs, some meritorious, some hereditary. Even though tribes had chiefs, their internals remained egalitarian, especially with regard to economic matters. However, the tribes themselves were not always equal, with some dominating others.

In time, some tribes coalesced, but each tribe of the supergroup retained its respective chief. In any group, someone has to be boss, and a group of tribes is no different. The chief of one of the tribes became the *capo dei capi,* the chief of chiefs. We advanced up the organizational chain from kin-related clans and tribes to chiefdoms and kingdoms. Each step improved security and, in tandem, increased population density, central organization, and political hierarchization. The key point is that even with all this organization, each tribe maintained its identity and chief.

Up to this point, we were tribal. Individuals identified with their local tribes, not with the tribe of the *capo dei capi.* We arrived at the portal of a second revolution. This was the creation of cities, the Urban Revolution, and with it, political organization changed. It went from a grouping of several semiautonomous tribes as the unit of organization to the state, where the individual was the organizing unit. In taking the step to where individuals identified with the new state and not with their old tribes, we became civilized. Civilized people are not tribal, and tribal people are not civilized. This will be our dividing line.

The foregoing is not a usual definition of civilization; the usual definitions do not serve our purpose. We need to make one specific for our needs. Our custom definition is more complex than just stated, and we will spend time fleshing it out. In general, the uncivilized identify with a local group and accept its authority. The civilized attach their identity to the broader state, whose absolute authority they must follow. For now, consider the subsistence farmer living in a rural, remote village as uncivilized and the specialized laborer in the city as civilized.

When our "out of Africa" ancestors arrived at what was to become Japan, Australia, Spain, or Peru, they did not encounter equal starting material in the form of resources. Some destinations, especially those in Eurasia, offered many plant and animal species for domestication. For these classes of resources, the Americas and

Africa had lesser variety, and Australia had very little. There were other differences, but the main ones relate to plants and animals. The ease of making the Neolithic transition was dependent on what geography or the environment provided as starting resources. Herein lies Diamond's main point.

However, just making the Neolithic transition was no guarantee of further advancement to the highest levels of human social development and organization, where we find the state and civilization. Some, like the Yanomamö of the Amazon, merely got to a toehold stage by combining subsistence agriculture with hunting-gathering. They have yet to progress further. Likewise, some pastoralists, such as the Bedouin, have never progressed very far up the organizational ranks. Even under the authority of the Ottoman state, their tribal organization remained untouched.[15]

It is not always clear why this or that group made significant social and political progress, or not, but what is clear is that those who made it all the way to "civilization" had access to adequate resources. Almost by definition, where civilizations arose, there were sufficient resources.

Civilization's reasons. There are two ways to look at "civilization." One is to view civilization as an entity such as found in early Egypt, the Incan Empire, or the Roman Empire, and we will refer to these entities; they have explanatory utility. The second, and less obvious, view is that civilization is a process that entails changing from rural, agrarian, and tribal ways to those associated with urbanization and the formation of the state. This is a key topic in our narrative.

There are two reasons for the coupling of civilization to adequate resources, one obvious, the other not. The rise of civilization is associated with cities, which equate to a high population density. With many people per square kilometer, there is a requirement for a high level of calorie production, and this translates to a certain density of resources. Civilization, along with urban development, was dependent on resource availability, and this is what environmental determinism explains.

The second, nonobvious reason relates to the formation of the state. It is with the city that the modern state arose. The state centralizes all the services and functions that until then the tribe had performed. States have bureaucrats, tax collectors, standing armies, nobles, royal families, and all sorts of other nonproducers for which the state must effectively provide food and shelter. For this, states need money or its equivalent, and its only source is from the people, the producers. Producers need resources in excess of their sustenance requirements so they can pay their fair share and still survive.

A state can only tax production in excess of the needs of the producers; it can only squeeze the orange so hard. For a state to form, the amount of available resources must enable a level of taxable production sufficient to support the state's needs. Simply stated, inadequate resources prevent state formation and the associated rise of civilization.

It is in the reason for becoming civilized or not where environmental determinism has its role. However, as evidenced by those who are civilized and yet remain poor, there is more. In developing our story, we will find that differences in the evolved culture of people and not their associated geography best explain these results; contrary to Diamond's thesis, people and their culture are the reason.

One key element that environmental determinism misses is the change in human nature with the advent of the state and civilization. The seemingly innocuous transition from identifying with the tribe to identifying with the state is one of humankind's most important transitions, one leading to far-reaching consequences. We are going to spend considerable time examining the nature of this transition and its consequences.

Geography's umbrella covers several elements, some more important than others. The most important is the availability of domesticable plants and animals. Animals provide not only protein but also labor and transportation. Provided sufficient domesticable plants and animals, most humans were able to make the transition of the Neolithic Revolution from nomadic hunting-gathering to sedentary agriculture. In addition to animals and plants, the list of geographic elements includes continental axis orientation, climate, access to water transport, and physical barriers to communication between populations.

The ease of movement of people and their ideas between populations was important for human development, with greater ease promoting more development. The nature of culture is that it copies; inventing the wheel only requires a single instance. Learning from others is only possible if others are reachable. In a way, a cultivated plant or its seed is akin to a cultural solution to a human problem. Once domesticated, it can become the next big thing in food anywhere it can be grown; it just has to get there.

There are significant differences between continents in their impediments to communication. They all have a different arrangement of communication-hindering mountains, rivers, deserts, and oceans. In Eurasia, there is the blocking continuum of the Himalayas, the Tibetan Plateau, and the adjoining Taklamakan Desert. Getting an idea or a seed from China to Europe was more than difficult. The Silk Road did not open until the Han Dynasty, about 200 BCE. By that time, China had been China for approximately two thousand years, with the Xia Dynasty having started in 2070 BCE.

Africa has the Sahara Desert blocking communication between Eurasia and sub-Saharan Africa. In the Americas, the big barriers are the Andes and the desert of northern Mexico and the southwestern United States. Of course, there are the oceans separating Eurasia and Africa from the Americas and Australia. On a macro scale, getting around was not simple.

If barriers to communication had not existed, we would all be the same, or at least more similar than we currently are. The free flow of ideas and genes would have resulted in a homogeneous human species, both genetically and culturally. We would have little diversity, and from the perspective of biology, that would be bad. Diversity is the foundation of a species's robustness for survival. With diversity, any subgroup or fraction can be eliminated and the species can continue. Diversity helps guarantee our survival as a species. Our differences attest to the existence of barriers to communication. We are different, and that is good for our long-term survival.

There is a secondary, very important consequence to environmental determinism. It is disease, which is probably only second behind plants and animals as a reason for success. People migrating to lands with domesticable animals eventually got as a lucky reward animal protein, labor, and transportation. They also got sick, and from the perspective of "survival of the fittest," getting sick was good.

Zoonosis is an infectious disease transmitted from animals to humans. Of the 1,415 pathogens known to infect humans, 61 percent are zoonotic.[16] The list includes cholera, cowpox, and plague; some believe that diseases like measles came from a mutant form of an animal virus. Populations exposed to disease develop resistance to that disease; exposure begets resistance. Those never exposed do not develop resistance and are at risk of catastrophic consequences when they are.

As long as people remain where they are born, there is no consequence to differential resistance to disease. However, man, being a curious creature, started to visit distant lands toward the end of the fifteenth century, the Age of Discovery. The results were monumentally disastrous for the discovered, host societies. When encounters occurred between people from domesticated, animal-rich environments like Eurasia and those from impoverished environments like the Americas, the latter got sick and died. Moreover, they died off in huge percentages. There were some reverse-direction diseases, such as syphilis, but they were minor in consequence.

When it came time to travel to new lands, especially to conquer, having locally nonexistent germs in your quiver paid off. They provided significant, unintended dividends to the conquerors. In my view, germs played a more important role than guns or steel when it came to who conquered whom.

Diamond's corollary. There is a corollary in Diamond's summary: it is that the different outcomes we observe between people have nothing to do with biological differences. For this, he provides only a little supporting biology. This aspect of his thesis appears to be more a statement of wishful ideology than fact. It is his agenda, an aspect sharply expressed by Nicholas Wade:

It is driven by ideology, not science. The pretty arguments about the availability of domesticatable species or the spreads of disease are not dispassionate assessments of fact but are harnessed to Diamond's galloping horse of geographic determinism, itself designed to drag the reader away from the idea that genes and evolution might have played any part in recent human history.[17]

Diamond's support of his corollary is in the form of an opening gambit in his prologue. He posits, based on personal observation, that the people of Papua New Guinea are more intelligent than Europeans are. His conclusion is that because they are more intelligent, any differences in outcome could not possibly be due to any supposed intellectual superiority of Europeans.

He argues that differences in social environment and educational opportunities confound making IQ comparisons between populations. This is true; properly, differences in IQ are only valid for within-population comparisons, where the environment and education are similar. This confounds comparisons of IQ between disparate populations. Confounding does not preclude differences. It only affects detecting them and has nothing to do with real differences being present or not.

He does offer two reasons for Papua New Guineans being more intelligent than Europeans or Americans. I do not want to add noise to our discussion, so I will refrain from analyzing his argument. However, I do want to present a countervailing perspective, one that is consistent with his expressed observations.

It is his subjective observation that Papua New Guineans are more intelligent than Europeans and Americans. We can define intelligence as the cognitive ability aiding survival and reproduction (biological fitness), and in an evolutionary sense, more intelligent people have greater fitness. Importantly, intelligence has a specificity. The intelligence we are talking about is specific for the evolutionary environment, and we can model this specificity by how a population makes a living. Foragers have a different evolutionary environment than slash-and-burn subsistence agriculturists, who in turn have a different one than those living in a modern, advanced economy. There is every reason to believe that the specific intelligence for evolutionary success in each of these three scenarios is different.

The Neolithic Revolution is when we changed our mode of making a living from hunting-gathering to agriculture, and this occurred approximately twelve thousand years ago. From that point, populations went through a series of changes in social and political organization. The progression was from egalitarian, kin-related bands and tribes up the hierarchal ladder of chiefdom to kingdom to state. Depending on where we lived, our means of making a living also changed with social and political organization.

It has been approximately two millennia since the ancestors of current-day Americans and Europeans last lived like the current Papua New Guineans. Over this period, the Papua New Guineans have remained the same. This translates to the Papua New Guineans spending two thousand more years in their current evolutionary environment than the ancestors of the people we refer to as populating the West. Consequently, their intelligence for their specific environment must be further advanced than that of Europeans and Americans. Given the specificity of environment and evolutionary time, they should have a higher specific intelligence than Europeans and Americans. They should be more intelligent, and on this Diamond is perfectly correct.

This is a red herring. The question that both Diamond and we are addressing is why some have better economic outcomes than others do. We need to address intelligence from this perspective of economic outcomes. For this, we need to address specific intelligence, intelligence that is specific for making a living in the modern world.

While the Papua New Guineans were spending an additional two thousand years sharpening their intelligence for their way of making a living, the ancestors of modern Americans and Europeans were sharpening theirs for the modern economy. Common sense and evolutionary biology both predict that when we use metrics of modern economies for comparative purposes, Americans and Europeans should fare far better than Papua New Guineans.

The World Bank 2013 data for per capita GDP for Papua New Guinea is $2,642 and for the United States is $52,980, or a twenty-fold difference. However, if we base our comparative metric not on the U.S. dollar but on how Papua New Guineans make a living, the difference should be the opposite. In a sense, the Papua New Guinean economy can be considered to be based their staple: yams. A browse through Wikipedia shows that Papua New Guinea produces 52 metric tons of yams per thousand inhabitants, while sweet potato (a tuber similar to yam) production in the United States is 3.2 metric tons per thousand, a sixteenfold difference. This is admittedly a simpleminded comparison, but it makes the point.

Diamond's flaw. Even though Diamond's thesis regarding the importance of people in the wealth and poverty of nations is lacking, we need to examine the other half of his thesis regarding environmental determinism itself. His theory has utility at one level and absolutely none at another.

It is at the granularity of continents where his thesis finds value. It explains why some populations from geographically advantaged Eurasia have historically fared much better than have those from continents not as well resourced.

Its critical failing is its inability to explain differences within continents, especially Eurasia. This is where we need to explain why England and Hong Kong are rich while India remains poor, or why Orthodox Christians have not done nearly as well as Protestants, or why Laotian Buddhists and Islamic Arabs are as poor as resource-poor sub-Saharan Africans. After all, they all evolved and lived on the same resource-rich continent and yet have markedly different outcomes. Because environmental determinism lacks explanatory power for the different economic outcomes within Eurasia, could differences between people offer the explanation?

CHAPTER 2

Laying the Foundation

⨉⨉⨉

We start our journey with a disagreement, one at the soul of this book. From there we will march toward a positive world, where dogma and political correctness find no solace or habitat—the world of science and rationality.

Paradigm of Psychic Unity

Anthropologists posit a dogma of modern liberal theology that we are all equal; we have not changed in the fifty thousand or so years since our migration out of Africa. Except for trivial physiological changes, we are now as we were then. The differences we observe in outcomes between people cannot be due to biology but rather must be due to other factors, such as the environment. People are not to blame or ever to be blamed; expressing any such sentiment is forbidden.

This paradigm of equality is well expressed by professor of archeology Lawrence H. Keeley in *War before Civilization: The Myth of the Peaceful Savage*:

> Anthropologists in this century have long argued for the "psychic unity" of humankind; in other words, all members of our species have within rather narrow limits of variation the same basic physiology, psychology, and intellect. This concept does not exclude individual variations in temperament or even the various components of intellect, but finds that such variations have no value in explaining

social or cultural differences between groups. It is not accidental that the descendants of illiterate villagers from various "backward" parts of the world, and a variety of racial backgrounds, have become Nobel Prize–winning scientists, mathematicians, and fiction writers using languages very different from those spoken by their ancestors. Anthropologists have long recognized that the many and profound differences in technology, behavior, political organization, and values found among societies and cultures can be best explained by reference to ecology, history, and other material and social factors. Thus, with a few rare exceptions, anthropologists argue with one another only about the relative importance of these nongenetic factors in explaining cultural variety and cultural evolution. This attitude reflects not just the antiracist tenor of the twentieth century, but also the accumulated facts and especially the experiences of ethnographers. Human psychic unity is not just a theory but a fact, one that can be demonstrated even in a survey of so dark a topic as war. The fact that despite our universal distaste we do "arrive where we started"—that is, at the blunt ugliness of war—unfortunately represents one of the clearest expressions of our shared psychology. Our common humanity, viewed realistically, can be as much a source of despair as hope.[1]

I disagree; we do not all have the "same basic physiology, psychology, and intellect," and developing a Darwinian rationale for my argument is this book's *raison d'être*. Answering the question of why some nations are rich and others are poor provides context.

Before presenting my case for disagreement, I want to speak to Keeley's manifesto. There are two points. The first is the arrogance of certainty contained in the statement "human psychic unity is not just a theory but a fact." This is dogma, dogma without tolerance, dogma not deserving a seat at the table of civilized discourse.

The second point relates to his example of Nobel Prize winners as evidence of "psychic unity." His statement about Nobel Prize winners is true. However, like much discourse today, it is misleading and, in reality, false. Yes, if we go back far enough, we are all "descendants of illiterate villagers from various 'backward' parts of the world." The statement is a red herring.

A reasoned look at Nobel laureates in table 2.1 suggests a conclusion at complete discordance with the one Keeley is promoting.

The data used in the construction of the table come from Wikipedia. The West comprises the nations of Western Europe and its

TABLE 2.1   Analysis of Nobel Prizes

Nobel Prize	Total Individual Winners	Won by the West (%)	Won by Ashkenazi Jews (%)
Peace	103	65	9
Literature	110	73	12
Chemistry	166	89	21
Physics	203	88	26
Economics	80	91	33
Medicine	204	93	27

derivatives (the United States, Canada, New Zealand, and Australia). It has a population of about 850 million, while the population of Ashkenazi Jews is approximately 11 million.

The Nobel Prize can be divided into two classes, one requiring analytical skills (Nobel Prizes in Chemistry, Physics, Economic Sciences, and Physiology or Medicine) and another not (the Nobel Peace Prize and the Nobel Prize in Literature). If psychic unity were a fact, there would be no difference between the West and the Rest. There is.

We can use Peace and Literature for normalizing between populations; the normalizing assumption is that for prizes other than Peace and Literature, the performance of the West should be neither better nor worse than the Rest. That is not the case; the West has a 22.25 percent edge, whereas "psychic unity" would predict no difference.[2]

The nail in the casket of "psychic unity" becomes set with an analysis of the performance of Ashkenazi Jews. Based on population size, they have won twenty times more Nobel Prizes across all prize classes than would be predicted. This excess of achievement ranges from eleven times for Peace to twenty-eight times for Economic Sciences. By the end of this book, we will have some understanding of the why of "psychic diversity," the actual nature of humankind.

Evidenced by our Nobel analysis, "psychic unity" is a false paradigm. However, that may not have always been the case. Once upon a time, when we were chimpanzee-like in Christopher Boehm's ancient forest[3] and living despotic, nonegalitarian lives, we probably had "psychic unity." Sometime between that distant then and now, we changed. Differences evolved, especially after our "out of Africa" migration, when, with dispersion, the Smiths could not keep up in an evolutionary sense with the Mings. As populations separate geographically, the genetic interchange that maintains genetic similarity decreases; the same holds true for cultural similarities. Geographic distance fosters divergence of genetic and cultural similarities.

Back then, we were just about like every other animal in the forest, with little portent as to whom we would become, the king of the jungle, forest, savannah, desert, oceans, and lakes—everything. We probably had as much culture as our primate cousins, the chimpanzees—not very much. The critical transition is that we, not they, evolved culture. This became our most important tool ever, because with this newly acquired and unique capability, we could now change our evolutionary environment. We did not know it then, and many still do not, but with culture, we could control our evolutionary destiny.

Obtaining culture resulted in the obvious of making clothes and cooking food. Most importantly, it included changing the environment of our natural selection. Culture gave us the ability to change our niches and, in so doing, change the direction of our evolution. Populations could and did differentially change their niches, and different niches led to different evolutionary results, both in physiology and in psychology.

This is an important concept, one leading to the refutation of the popular contention that we have not significantly changed in the past fifty thousand years. This is the argument that our only evolutionary changes have been nondeterministic, cultural ones. The dictum that we are all genetically the same as we were before leaving Africa is simply not correct.

That we did or did not acquire race-specific genetic mutations, other than the trivial ones relating to climate adaptation, is not in this debate. Adaptive genetic mutations becoming fixed in a population is only one mechanism for creating genetic differences; there is another. Various populations, through constructing their own niches with culture, have altered evolutionary pressures. The result is that certain gene variants (alleles), which exist in all large populations, increase in frequency and become dominate in their respective populations. Different populations have the same set of available genes but, owing to local niche construction, evolve different dominant ones for the population. At a minimum, we are genetically different because we have different gene frequencies, with some of these differences reflected in different behavioral traits.

When some of us took up herding cattle, the evolution of lactose tolerance past the weaning stage and into adulthood occurred. Being able to drink and tolerate milk was adaptive; milk drinkers had higher survival rates, a key element in Darwinian or natural selection. Milk provided not only protein but also a clean source of liquid, an important issue when water was probably contaminated.

The genetic mutations responsible for adult lactose tolerance are evolutionary footprints. They evidence culture modifying the environment for evolution. Culture by herding cows did not create the alleles; it only changed the environment. The changed environment led to natural selection selecting for alleles promoting adult lactose tolerance, enabling adults to digest milk.

In the case of adult lactose tolerance, our physiology changed due to culture changing the evolutionary environment. Certainly culture has also been responsible for some of the evolution of our psychology or human behavior; we just do not have the obvious footprints as evidence. The evidence might be present, but the problem is that the genetic component of any behavior or trait probably consists of hundreds, if not thousands, of alleles. We are just now beginning to be able to identify such groups of alleles.

The American economist, behavioral scientist, and educator Herbert Gintis provides a clarifying quotation:

> Gene–culture coevolution is the application of niche construction reasoning to the human species, recognizing that both genes and culture are subject to similar dynamics, and human society is a cultural construction that provides the environment for fitness-enhancing genetic changes in individuals.[4]

Individuals express their culture, but culture is properly a property of the group, not the individual. Thankfully, for *gourmets* and *bons vivants*, there really is a French culture. Over the course of human history, there have been thousands of unique human groups, and even today, ancient groups persist well into the presence of numerically superior groups. They remain coherent and have not been absorbed or exterminated. The Nakhi (Naxi) of China and Parsee of India are examples. In the past, some groups, such as the Ionians and Aeolians, combined to form a new identity, the Mycenaean Greeks. There is also fissioning, a process creating new groups. Of all the human groups that ever existed, many, probably most, no longer exist, and whatever cultural uniqueness they possessed is gone.

If culture never evolved, perhaps we would still all be the same and "psychic unity" a reality. Nevertheless, it did, resulting in our "psychic diversity." Our relatively recent history is one consisting of many groups, each having unique culture and, with each culture, creating different environments. Some have led to different selection pressures that in turn resulted in different evolutionary outcomes. Innumerable groups, various cultures, different evolutionary environments, all leading to different evolutionary outcomes—why would anyone consider "psychic unity" to be a fact? Perhaps certain belief systems demand that conclusion.

Our Topic

For our topic, we are not concerned with the inequality of physiology, where we find differences between populations in skin color, hair type, and susceptibility to certain diseases. Our concern is with the differences in behavior and traits between populations. Our interest is not in the empirical description of psychic diversity; it is in elucidating the process of how it came to be. Furthermore, it is in understanding how these differences in process translate into the observed inequality of economic performance and outcomes between nations.

Our premise is that differences have reasons and that the Darwinian theory of evolution offers the best explanation. Evolution is not just about genes; it includes culture. Natural selection is a bottom-line process. It has no knowledge of the etiology of a trait under selection and does not care. It could be genes, environment, culture, or any of the possible interactions. It bases selection on the result of all those forces.

Using principles of Darwinian Theory, we are going to develop a model that explains how our several populations have arrived at their current condition of "psychic diversity" or inequality. Our assumption is that, if we are different, we must have followed different evolutionary pathways, with different pathways concluding in different ends, and these ends are the differences we observe. Understanding the process leading to these differences should provide the basis for understanding why we are different, a far more insightful question than how. We will develop reasons that are understandable, rational, and based on science, even if some are speculative.

Not all the science is settled and some ideas are conjectural. However, this incompleteness of settled science does not preclude making a holistic model; this book is a working hypothesis. At the end, I hope you find the model to be consistent with your observations. The model should rationally predict reality.

Narrative's Nature

The first part of this book contains some of the parts and pieces needed for developing our account, and this is where we just now presented the core question. Here we will spend some time on the consequences of economic inequality, establishing the base problem within a broader set. So those lacking expertise in biological sciences can follow the logic, we will present some evolutionary principles using, it is hoped, an unintimidating, broad brush.

For our story, we need to classify populations with a view of making comparisons. Additionally, because our interest is in why populations differ, we need to approach any classification scheme with the view of "common evolutionary history."

Choosing a basis of classification is an important topic to which we will return later. When we do, we will show that we can satisfy both requirements with a schema primarily based on religion, amended by geography.

Once we have a system for classification, we need data for analysis. We will get some and, in the process, create metrics demonstrating economic differences. With numbers, we will have something tangible about which to talk.

The middle part of this book is its core. Here we develop our model, explaining why people are different. The process of model creation was a journey of discovery. There were no *a priori* building blocks; they were all developed along the journey. It was a path entailing a layer-by-layer uncovering of the clues to human development. I could present the model in the usual top-down fashion. However, I believe that if readers follow the same bottom-up path leading to discovery, they will arrive at the same logical end. I hope that this approach will lead to a higher level of agreement than otherwise.

In the last part, we apply the model to our current world. Here we examine the various civilizations within the context of the model. We will see that our model explains the differences between populations that led some to be rich while others remained poor.

How Different Can We Be?

The basic elements of a computer program are its zeros and ones, and their sequence defines a program's information. We, you and I, have such a program code defining who we are. It is our genome. Conceptually, it is the sum of our genes.

Our DNA, which defines our genome, is not zeros and ones; it consists of a double strand of nucleotide sequences organized as base pairs. The four types of DNA nucleotides are the molecules adenine (A), cytosine (C), guanine (G), and thymine (T), and a section of a single strand might be

$$T\ G\ C\ A\ A\ G\ T\ C\ C\ G\ A\ T$$

DNA is a double strand, the famous double helix. The chemistry is such that a C in one strand is always attached to a G in the parallel one, and an A in one is correspondingly attached to a T. If we use the preceding single-strand example, the resulting double strand would be as follows:

T G C A A G T C C G A T

A C G T T C A G G C T A

This shows the base pairing. Given the nature of these specific chemical bonds, the resultant form is a double helix. In addition, the rules for pairing define the method for replicating DNA molecules, an important process occurring in our cells.

The metric defining the size of a computer program is its number of bits. A genome has a similar metric, the number of base pairs. The human genome has approximately 3.2 billion base pairs distributed unevenly between twenty-two chromosome pairs plus two sex-determining chromosomes, X and Y. We inherit twenty-three chromosomes from our fathers and twenty-three from our mothers.

If my genetic code were identical to yours, we would be identical, but they are not. There is variance between our codes, where *single nucleotide polymorphism,* or SNP (pronounced "snip"), is the term for this variance. A SNP is a DNA sequence variation occurring commonly within a population. The term *allele* is for one of a number of alternative forms of the same gene, a gene variant. SNP and allele are approximately the same.

A section of your DNA might have the sequence

T G C A **A** G T C C G A T

and mine could be

T G C A **C** G T C C G A T

Humans have about one SNP per six hundred base pairs. This value quantifies the maximum size of our genetic difference, which is only 1/600, or 0.167 percent. Expressed another way, we are 99.833

percent the same, but a miniscule 0.167 percent of a very large 3.2 billion base pairs is still a substantial 5,344,000.

Some have argued that this miniscule difference is evidence that we are genetically all the same, but this is really a foolish argument. Both importance and our interest are in the quality, not quantity, of differences. Our concern is with important psychological difference, be it altruism, despotic behavior, or intelligence. A simpleminded numbers game is interesting but adds little to our discussion.

Some have used the numbers game to proclaim that there are no genetic differences between races or ethnic groups; they are pronounced as human constructs. It is true that no single SNP can be used reliably to differentiate between groups. However, what is sometimes left out of the argument is that when an analysis uses several SNPs, the use of either cluster analysis or principal components analysis approaches 100 percent reliability in assigning individuals to populations or groups. This is the technique used by the company 23andMe for purposes of determining what percentage of a client's DNA is from various populations around the world. Not only is your DNA a statement of your unique identity but it tells us where your ancestors lived.

Basics of Change

In a previous section, "Paradigm of Psychic Unity," we brought into our discussion the concept of Darwinian or natural selection. Before we go too much further, we need a rudimentary understanding of how this functions; it is critical to the model we are going to develop:

> Natural selection is a causal process. Typically (but not necessarily) there is variation among organisms within a reproducing population. Oftentimes (but not always) this variation is (to some degree) heritable. When this variation is causally connected to differential ability to survive and reproduce, differential reproduction will probably ensue. . . .
>
> Why is it that some variants leave more offspring than others? In those cases we label natural selection, it is because those variants are better adapted, or are fitter than their competitors. Thus we can

define natural selection as follows: Natural selection is differential reproduction due to differential fitness (or differential adaptedness) within a common selective environment. . . .

As a causal theory natural selection locates the causally relevant differences that lead to differential reproduction. These differences are differences in organisms' fitness to their environment. Or, more fully, they are differences in various organismic capacities to survive and reproduce in their environment. When these differences in capacities are heritable, then evolution will (usually) ensue.[5]

"Variant" is the next piece of the puzzle. Organisms have genes defining them but what natural selection sees is the expression of those genes in the environment. If genes define a genotype then the expression of the genotype in the environment is the phenotype.

phenotype → genotype + environment + interactions

It is essential to distinguish the descriptors of the organism, its genotype and phenotype, from the material objects that are being described. The genotype is the descriptor of the genome which is the set of physical DNA molecules inherited from the organism's parents. The phenotype is the descriptor of the phenome, the manifest physical properties of the organism, its physiology, morphology and behavior.[6]

Stated less formally, the genotype in the preceding expression is simply the genes in an organism's DNA responsible for a certain trait and the phenotype is the expression of that trait in the actual environment. Natural selection can only select based on the real-world expression of the trait or what it "sees"; that is why we separate phenotype from genotype.

There is another requirement for natural selection: variation. If all the individuals in a population have the same expression for a trait under selection, there would be no basis for selection; it would not occur. Selection requires variation; else, evolution has no basis on which to choose. This will become important later in the book when we discuss egalitarian societies. In such societies, variation between individuals is minimal, affecting natural selection.

Natural selection is all about survival and reproduction, the last being the obvious step needed to get heritable traits passed on to the next generation. Those heritable traits that differentially increase survival and reproduction are termed *adaptive*, and individuals possessing such traits are termed *fitter*. The important point is that with this process, adaptive traits increase in the population, rendering the population better suited for its habitat.

A key operative word in the foregoing is *differential*. This infers a competition where some survive and reproduce and others do not. This is correct; evolution is a contest with winners and losers, and the winners are those getting the greater number of copies of their genes into following generations. We will later see that males receiving societal rewards of multiple mates are biologically fitter than males constrained to just one. They will have more progeny than others will, establishing them as evolution's winners.

Culture and the Changing of Populations

The preceding ideas are all well and fine for everyone else in the biological world, but we are special. We have culture, so much that we can change our evolutionary environment; we can alter the environment for natural selection. The earlier story about adult lactose tolerance is an example.

For humans, the expression for phenotype should be

$$\text{phenotype} \rightarrow \text{genotype} + \text{natural environment} \\ + \text{cultural environment} + \text{interactions}$$

Our phenotypic traits come from not only alleles but also the heritable portion of culture and gene—cultural interactions. The main point is that with natural selection, the heritable portions of selected traits, whether from genes, culture, or both, will have their frequencies increase in the population. This is the way populations become different.

Even though culture is part of the human phenotype, it is different from deterministic genes; we do not inherit culture in the same manner. Its mode of inheritance is learning. Look closely at the behavior of parent and child; it is obvious. Parents who say please and thank you have children doing the same. This is not trivia. It is the learning of societal rules for behavior; it is the acquiring of culture. The evolved social norms for bringing a drink when "please" is the antecedent, are different than when the drink is brought in response to a command. This behavior brings cohesion and strength to the clan, making them collectively stronger for it.

We can find another example in the simple act of visiting a foreign country, especially one with different rules for social behavior. Many, if not most, travelers encountering this situation instinctively modify their behavior to mimic the behaviors of the host country. In Japan, the Westerner makes the slight bow on greeting, and in France, all say, with a charming but curious lilt to the inflection, "*Bon jour*." We want to blend in, to belong. To do so, we observe, learn, change, and adapt; as a result, everyone, host and guest, is more comfortable for it.

Culture plays a role in behavior. Japanese bow, Americans shake hands, and the French shopkeeper invariably says "*bon jour*." These behaviors, and thousands more like them, pass on generation after generation, becoming a unit of the heritable portion of culture. However, what is important is the fact that culture affects our evolutionary paths by altering the phenotype seen by natural selection. The following quotation gives substance to this concept:

> Culture is a system for the inheritance of acquired variation. What individuals learn for themselves by hard effort others often imitate, typically at much less cost. . . .
>
> So long as people are behaving differently because of cultural or genetic factors and succeeding differentially because of those factors, selection is operating.[7]

This last quotation expresses our bottom-line conclusion.

Heritable Traits

Natural selection selects the phenotype, but what passes to the next generation is not the phenotype. What passes are the heritable factors determining the phenotype, namely, the genes of the genotype and the heritable portion of culture influencing its environment. The point for this section is that what passes to the next generation comprises more than just genes; it includes the heritable cultural traits defining and forming the phenotype.

This is an important concept, and accordingly, we will use a quote from a recent paper for clarity. Note that the authors are discussing the transmission to and the inheritance of information by the next generation. Evolution is about changing and improving the information available to future generations for their survival and reproduction. The important concept is that evolutionary information has more than just genes as a source; there is culture:

> We offer the concept of "inclusive heritability" to identify and unify the various forms of information that are inherited, both genetically and non-genetically.
>
> One form of non-genetic information is animal culture, which may affect evolution via several processes. Culture is produced by processes of non-genetic information transmission that are analogous to DNA transmission. Just as genetic mutations are inherited, socially learned innovations can also be inherited. However, whereas genetic information can be transmitted only vertically (from parents to offspring), cultural traits have the unique properties of also being transmitted horizontally (among members of a generation) and obliquely (among unrelated individuals of different generations). Learned innovations can spread through a population within a single generation and affect behavioral adaptations. Variation in cultural information may thus contribute to phenotypic variance. Selection can thus act on variation that is produced by both genetic and non-genetic inherited information.[8]

In our narrative, we will often use a term like "heritable traits from alleles and the heritable portion of culture will increase in the population." Sometimes we will keep it short with just "heritable traits will increase." The foregoing is what we mean.

Nature or Nurture?

The intertwining of genes (nature) and culture (nurture) along with their interactions makes it difficult to ascribe cause and effect. Separating nurture from nature is difficult at best. Fortunately, identical twins offer a clever method to gain insight to this question. At conception, they are genetically equal, but differences can and do creep in, especially during development. Furthermore, just being raised under the same roof offers no guarantee of environmental equality. Even with these caveats, they are highly similar.

Just like the majority of other children, most identical twins are raised in the same household. Some, however, are separated at birth, to be raised in different families. Now we have the same genes in the same environment versus the same genes in different environments. With this condition, we can make scientifically proper comparisons. We can use the variability of a trait between same and different environments to obtain an estimate of its genetic heritability, if any.

IQ literature has genetic heritability at about 50 percent, an estimate derived from studies of identical twins. A key point is that heritability is not a measure of amounts; it is only a measure of variability. Pointedly, the simple fact that this variability exists is proof that the trait under consideration has a genetic component. Studies of identical twins have demonstrated that traits such as personality and intelligence have genetic components. This is true even though we lack explicit knowledge of the identity of the specific alleles. Furthermore, it establishes the blank slate or *tabula rasa* paradigm as clearly false. We are born preassembled with our propensities, just like our appendages, already in place. Do not take my word for this; ask any parent with two or more children.

Polygenic Traits

Eye color, blood type, and lactose intolerance are examples of monogenic traits. Traits, such as intelligence, personality, sexuality, and height, are the expression of hundreds to thousands of genes, each with its allelic variations. These traits are polygenic. The difference between monogenic and polygenic should influence how we view evolution.

The earlier story of cattle raising that resulted in adults being able to digest milk exemplifies much of what many consider as evolution. Recall that when humans first started to domesticate cattle, adults could not digest milk. Being able to do so awaited a genetic mutation for the extended production of the enzyme lactase into adulthood. If that mutation had never occurred, adults today still would not be able to drink milk. We would remain lactose intolerant as adults and, importantly, I could not enjoy ice cream.

Fortunately for ice cream lovers, mutations for extended lactose tolerance occurred, and it happened independently in various places. It turns out that being able to digest milk is adaptive. Not only is milk a good source of protein but it is also a clean source of liquid in what was then an otherwise unsanitary environment. Being capable of digesting milk was adaptive, and those who raised cattle and had the allele out-survived and out-reproduced more successfully than those who did not. Like any adaptive allele, it spread throughout the population via natural selection. Its frequency increased in the population's gene pool.

The preceding discussion illustrates the common perception of how evolution occurs. It is where a mutation is adaptive and, through natural selection, increases its prevalence or frequency in the population and, in the process, replaces the older, less adaptive variant. Monogenic traits change this way.

Polygenic traits with their hundreds to thousands of genes, each with its various allelic forms, present another pathway to change. We can use intelligence for our example. Assume that changes in the environment, natural or man-made, result in changes in the selection pressures of natural selection for intelligence. Perhaps man-made changes resulted in a more complex environment, making greater intelligence strongly adaptive. Being polygenic, natural selection has almost innumerable allelic forms for selection, any one of which only very slightly increases intelligence. Evolution does not hang around, waiting for a fortuitous mutation; it immediately reacts to the change in selection pressure by selecting those alleles already present in the population that are best suited to responding to the change. It works with what is already there by rearranging the preexisting deck chairs. The result is that the frequency of no single allele increases markedly; however, the frequency of the sum of all the individual ones increasing intelligence does. The population becomes smarter.

This process of adaptation is applicable for polygenic traits, with some researchers believing that this has been the dominant process in recent human evolutionary history.[9] As we develop our narrative, we will see that this process could explain many of the differences observed between populations.

With a monogenic trait like eye color, eyes are blue or not. Polygenic traits, not being all or nothing, result in incremental differences in the expression of the trait. When a large number of alleles defines a trait, it has a continuous distribution of values over a given range, with values further from the population's mean being less common.

This property of incrementalism is important for a polygenic trait like intelligence. If gradations were not part of the deal, all the allelic cards would have to be precisely ordered to change a population from not so smart to very intelligent, an all-or-nothing condition. With gradations, all that is needed is for most of the cards to be in the deck and in play in order to be smart.

The nature of polygenic traits leads to a numbers game with interesting and important conclusions. For a trait like intelligence, large populations probably have in their gene pools the vast majority of alleles responsible for that trait. However, small populations might not. There is an important consequence to the foregoing. If one ethnic group has a significantly higher IQ mean than another, it is highly unlikely that it is due to a few differences in mutations or alleles. Consequently, they are not *a priori* due to genetic differences in race or ethnicity; only differential polygenic adaptations offer an explanation.

To learn if a population has the "right stuff," there is a simple test. If only a few of a population exhibit the high end of a polygenic trait, say, an IQ of 130, most of the alleles needed to express it probably exist in the population. If no one exhibits the trait at the high end, then we can say nothing and would have no knowledge if the alleles were present or not. We would have to devise another test in order to know.

What is important is in what we can say. The missing ingredient is the selective pressures from natural selection. There has to be a reason for natural selection to change allelic frequencies of a population. If pressure is not present, applicable allelic frequencies will not change, and neither will the population. Only if one ethnic population has a greater pressure for selection for IQ will it become smarter than the others will.

CHAPTER 3

Inequalities' Consequences

There are those who, before starting a mystery novel, will read the last few pages to find out how it ends. We can satisfy any such curiosity by looking at two observations found in Clark's *A Farewell to Alms* and, with them, arrive at our ending.

In the mid-eighteenth century, before the start of the Industrial Revolution, the world's population growth rate was not zero, but close—a paltry 0.6 percent. At this rate, the time needed for a population to double is approximately 115 years.

In some societies, medieval England and China being examples, rich men had more surviving children than poor men. This is expected. Children of the rich have better nutrition, especially in times of scarcity. They did not starve while the poor did. They also lived in conditions that were more sanitary, whereas the poor often lived several to a room, where "several" could include pigs, dogs, and chickens. Disease from lack of sanitation kept the poor population in check; it killed off their children. The rich were by no means clean, just less filthy. We could add to this mix other Malthusian population control factors, such as war and pestilence, but the first two make the point.

Having just these two observations, a population biologist could predict the following. Societies with differential living standards due to wealth inequality, social mobility, and zero or low population growth would have the frequency of any gene variants (alleles) promoting wealth creation increase in their populations. Consequently, over time, such populations would have an increase in their economic acumen, and they should become more economically prosperous than those not engaging this process will become. This result is what we observe in the world today.

We arrive at this conclusion by understanding that the children of rich men through downward mobility fill in the population spaces not filled by the children of the poor, a result required by the mathematics set by the condition of a zero growth rate. Importantly, when the children of the rich fill in the spaces below, they bring along their heritable traits, especially those leading to wealth creation.

There is more to this story; it is complex, interesting, and perhaps controversial. A controversial aspect is that if intelligence is heritable (it is), and if it aids in becoming rich (it does), then our example population should become more intelligent, at least in matters pertaining to wealth creation. Assuming the validity of the foregoing, wealth inequality has led to some populations becoming smarter than others are.

Some might find this conclusion disconcerting. It is more pleasant to hold the belief that we are all the same, each with equal capacity, and that any observed differences in outcomes are due to "different environments."

Being disconcerting is not my cause. My cause has an opening premise—there must be a reason for populations arriving at different economic outcomes. We are the result of Darwinian natural selection operating on the variance of individual and group traits. The origins of these traits are genes, culture, and gene–culture interactions.

Differences arise because we have different environments prompting the direction of natural selection. They result in different selection pressures, which in turn result in different outcomes.

The environmental factors that direct natural selection include not only the usual suspects of resources and climate but also human culture. With culture, we can modify our evolutionary environment. We have already discussed such an example: dairy farming leading to lactose tolerance. Yes, "different environments" are the ultimate cause of differences in outcomes, but the process is Darwinian, not some vague, undefined effect on people. Whatever their source, natural or man-made, different environments have resulted in different evolutionary products, the "us" of today, people with differences.

Understanding how we are different is important, but answering why provides considerably more insight. We should be able to use Darwinian evolutionary principles to derive the why of arriving at these differences, and this is what we are going to do. Additionally, solving problems like inequality requires knowing why more than knowing how. This is the strength underlying the use of Darwinian principles.

Nations are not just lines on a piece of paper denoting an enclosed area; they are people, some homogeneous and others heterogeneous. These people have all been on their respective evolutionary journeys, ones that that probably commenced with equality some long, long time ago. Along this journey, differences arose, some of which account for wealth inequality between populations.

We are going to develop an understanding of why these differences came to be. This will be a path of discovery where, layer by layer, we uncover the nature of the why of our differences. By journey's end, we will have a model of why we are; the how falls out in the process.

This book is primarily about the process by which people have become different. This could have been the book's solo topic—economics, nations, wealth, and poverty did not have to be part of this conversation. However, conjoining these issues into an interrelated topic is, in a cognitive sense, making the entirety of all issues clearer. It is an explanatory symbiosis.

Inequality and Darwinian Fitness

The story at the start of this chapter relates that wealth inequality in preindustrial England resulted in increased survival and reproduction for the rich. This observation by Clark is important because it implies that wealth can provide its possessor with biological fitness. This is a case where a property of culture—wealth—gives guidance to evolution's direction. It has the probable consequence of heritable traits that promote acquiring wealth increasing their frequency in the population.

The foregoing presents a sort of mixed metaphor. By improving living standards, wealth increases biological fitness, and therefore traits promoting wealth should increase in the population. If achieving wealth were solely due to heritable traits, there would not be a problem, and we would not need this section. However, some acquire wealth when it is passed from one generation to the next, and other than luck in selecting one's parents, wealth acquisition has nothing to do with any traits. This dual nature of wealth, earned versus unearned, confounds the model, making it an issue we need to understand.

The basic idea underlying Darwinian evolution is as follows:

> If there is random variation among the traits of organisms, and if some variant traits fortuitously confer advantages on the organisms that bear them, i.e., enhance their fitness, then those organisms will live to have more offspring, which in turn will bear the advantageous traits.[1]

There are two factors in the preceding quotation. One relates to heritable traits and the other to their consequences. As Clark's story shows, a consequence of wealth can be increased fitness. Unfortunately, wealth, while inheritable in a cultural sense, is not a heritable biological trait, an implied requirement in the quotation. This inconsistency needs resolution, and that is what we are about to provide. We are going to address the question, where does wealth fit in the scheme of Darwinian Theory?

The first point we need to make clear is that we are talking about differential wealth or wealth inequality. It is only when those with more wealth out-survive and out-reproduce more successfully than those with less that wealth has a role. Evolution is competitive, and Darwinian evolution is all about differential survival and reproduction. If everyone is equally wealthy, wealth has no differential consequence, and there is no resulting change in the population. Thus what we are talking about is wealth inequality, with the wealthy having greater fitness than the poor do.

(We should note that the actual increase in fitness comes from better living standards. Furthermore, wealth affects more than living standards. For example, scions of the rich usually have a lower probability of going to war, also resulting in an increase in survival. We are going to keep it simple and use only "improved living standards" in our discussion.)

That wealth inequality rewards the wealthier with increased fitness through better living standards is straightforward. It is in the source of wealth inequality where confusion enters. Inherited wealth is not a consequence of biological traits and, consequently, does not result in a change in the population. Only heritable traits affecting Darwinian fitness can do that.

The story is different when the source of wealth inequality is income inequality. Traits such as intelligence, tenacity, and willingness to accept delayed gratification have a role in economic success. There are others, but the main point is that these traits probably have a degree of heritability, and in societies where income inequality leads to wealth inequality, the prevalence of heritable traits promoting wealth should increase.

Heritable traits lead to income inequality, which in turn leads to wealth inequality. This means that they can result in increased fitness. We can view the proximal cause of the fitness increase as wealth inequality, but from the biological perspective, it is ultimately income inequality.

That wealthy individuals are biologically fitter in some environments and for certain societies is a very powerful concept. When income inequality results in fitness-enhancing wealth inequality, it can lead to some populations having superior biological consequences. It is a positive feedback system where income inequality increases fitness, which in turn causes an increase in the frequency of heritable traits leading to acquiring wealth. The population becomes wealthier, and importantly, there is an increase in the frequency of income-promoting traits among its individuals. People become smarter, become more tenacious, and have a greater willingness to accept delayed gratification. This is a mechanism for change.

Some might believe that this book advocates for the goodness of wealth or its creation, and that would be wrong. This book is not about value judgments. After all, we are all capable of making them for ourselves. All this book can do is provide some amount of rationality to the decision-making process.

This book is about biology, and biology has its own innate sense of good and bad. Any perceived advocacy for wealth is in the context of the imperatives of biology. Biology is all about getting genes into the next generation, and any process increasing that probability is good, at least from biology's perspective.

We can even go a step further and state that, from biology's perspective, individuals and populations possessing the greater number of fitness-enhancing traits are superior. Income inequality, but not wealth inequality due to inheritance, can fit into that category. We might not like the idea of considering one group or another as being superior, especially when the basis for the division is income inequality, but this is about biology, which cares not a whit about our opinions.

If we do not like biology's conclusions, we can reframe the argument to reference our respective cultures. We can decide what is good and bad, but in the process, we simply must recognize the reality that biology gets an irrevocable vote.

What is important to note is that we can use our culture to change biology's conclusions, and later, we will learn that in our past, we have used culture to do just that. We altered Mother Nature's intent when we fooled her by collectively repressing alpha male behavior. Never forget that our unique culture is a potent force in its own right.

The Industrial Revolution and Economic Inequality

In the mid-eighteenth century, a revolution started in England that was to lead to our current high level of economic inequality. This was the Industrial Revolution. The machines of the Industrial Revolution gave us factories, along with factory workers, but most importantly, they gave us efficiency of production. A few

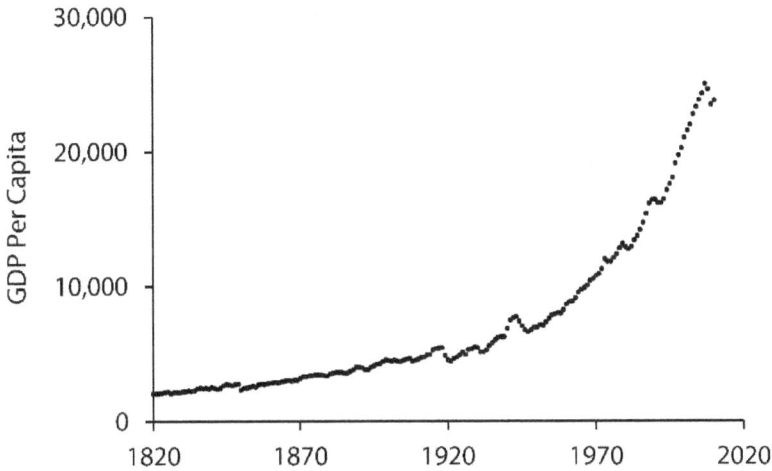

FIGURE 3.1   UK GDP Per Capita

could make what used to require the effort of many. Efficiency led to lower costs, which in turn led to increased demand. Efficiency led to wealth. Wealth in the form of capital in turn led to investment in more technology, offering further increases in efficiencies, and so on, in a giant feedback loop that continues today in similar form within developed nations.

Since the beginning of time, GDP per capita or its equivalent had hovered near zero. The following figures pick up this story in the early nineteenth century, when we first begin to see the effects of the Industrial Revolution. The sharpness in the change in the U.K. GDP per capita curve (figure 3.1) indicates that this engine of wealth creation started up very quickly. Where the Industrial Revolution took hold, economic output rose quickly due to continuously more efficient production, whether for textiles or cast iron. What was important was that poor farm laborers could move off the farm and into higher-paying occupations requiring more skill, provided they had that skill. The poor could become lower middle class, the lower middle class could become middle class, and so forth.

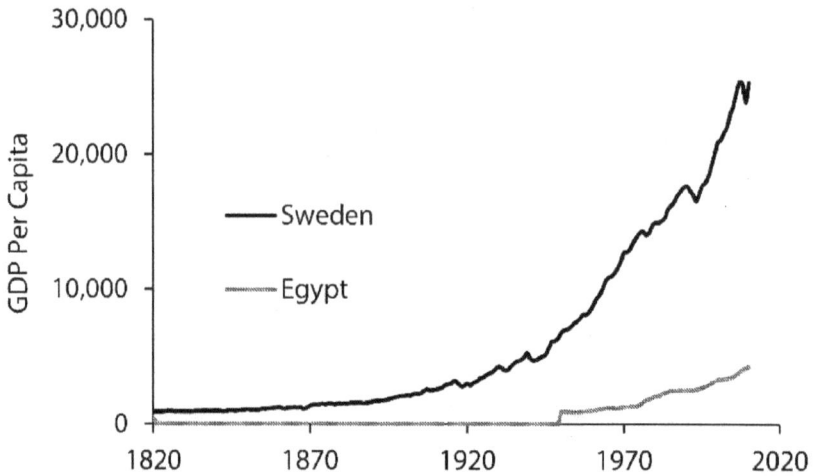

FIGURE 3.2   Sweden versus Egypt for GDP Per Capita

Not all nations grabbed the revolution's brass ring, but those who did have graphs similar to that of figure 3.1, the main difference being a later starting time. Figure 3.2 illustrates this point with Sweden not starting to take off until later than the United Kingdom. Figure 3.2 also shows Egypt, like the rest of the poor nations, never making the change. This divergence in economic growth between the brass ring grabbers and the brass ring droppers is the Great Divergence; it represents a major problem in this modern world and is a core issue of this book.

The Industrial Revolution certainly enabled some of the existing rich and well connected to become richer, but much more importantly, it created the demand for skilled workers, with skilled workers making more money than farm laborers. It created the path of upward mobility. The nations that grabbed the brass ring started to form societies that had more than just the few rich and the innumerable poor. It created all kinds of classes based on profession, with different professions resulting in differential incomes. There were now more class increments, such as upper middle class or lower middle class, between the poor and the rich. With economic mobility, the poor could move up. The rich could also move down.

This downward mobility is important to the development of our thesis. However, contrary to the belief of some ideologues, mobility within a system of economic inequality is not a zero-sum game. The rich are not becoming richer at the cost of the poor becoming poorer. The reality is that the sum of the wealth of all the individuals increases. Nowadays, the poor of the rich nations have more wealth and a much better quality of life than that of the middle class of many poor nations.

Even now, some 250 years after the start of the Industrial Revolution, many nations remain poor, whereas others have become wealthy. Rich nations like Norway and Singapore have per capita GDP in excess of $54,000, while the sub-Saharan nations of Zimbabwe and Democratic Republic of the Congo generate less than $560—a hundredfold difference.

Within some poor nations, some dominant minorities are relatively much wealthier than the majority of their neighbors. For example, in the Philippines, where the per capita GDP is only $4,380, people of Chinese descent comprise only about 1 percent of the population, but they control 60 percent of the private economy.[2]

In addition, within wealthy countries, there are relatively poor minorities. In the United States, the median income of African Americans households, at $33,321, is about 58 percent that for whites ($57,009) and only 48 percent that of Asian Americans ($68,636). In an almost separate universe, there still exist tribes having economies based on gifting or reciprocity. They are not the market base of the modern world. In these societies, wealth, in the sense of modern economies, is not an understandable concept.

The Existential Threat of the Poor

One of the first things taught in a course of physical oceanography is that the earth is an oblate spheroid, an erudite word for "not quite round." The second is that the average depth of the oceans is four thousand meters. In perspective, if the earth were a basketball that you were holding, you might notice that it was slightly damp.

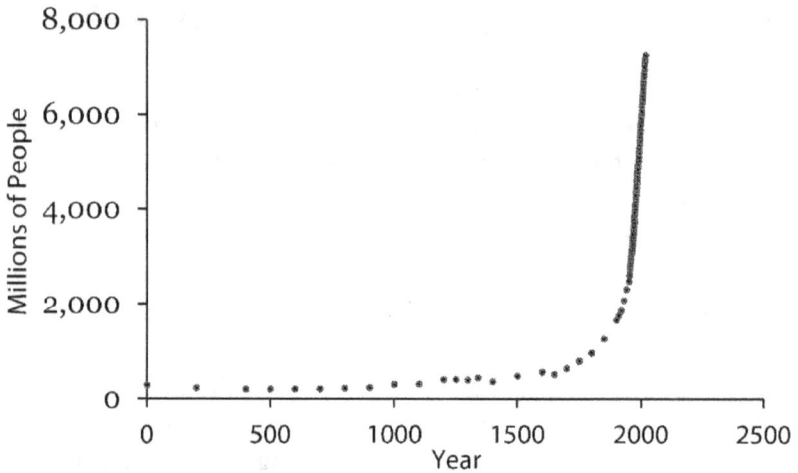

FIGURE 3.3   Growth of World Population

This "slightly damp" layer approximates our biosphere. It is thin, finite, and resource limited, a nature discordant with the graph of the growth of the world population in figure 3.3.

Reality cannot sustain the population growth that has occurred since the advent of the Industrial Revolution. Something must change, or our genes, both yours and mine, have a problematic future. I do not know the meaning of life, but the meaning of biology, its imperative, is getting our genes into the future.

It is not intuitive, but solving the problem of wealth inequality relates to solving the problem represented by the population growth shown in figure 3.3. As we will see, lifting poor nations up from poverty should cause the curve to bend and flatten out. Consequently, if the problem represented by poor nations persists, so does the existential threat of the population boom. Let me explain.

The heart of the argument is in the five-stage Demographic Transition Model shown in figure 3.4. Much of the data supporting the model came from developed countries like the United Kingdom, where historical data exist. Consequently, the applicability of the model to less developed countries is problematical.

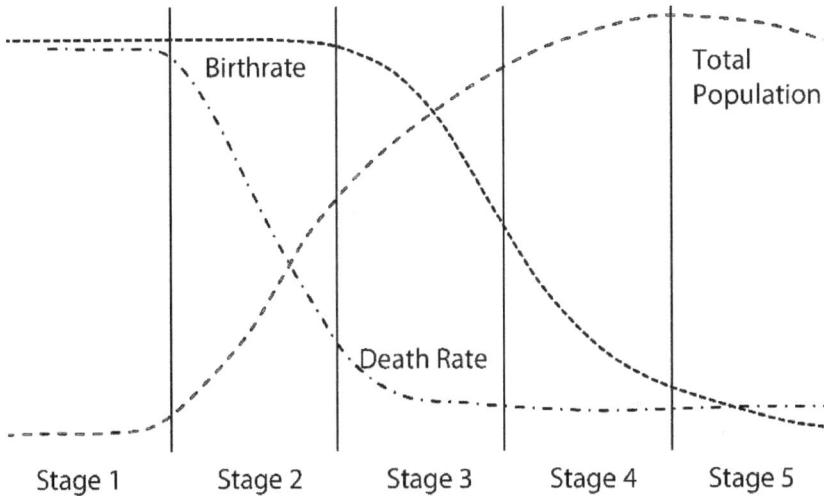

FIGURE 3.4   Demographic Transition Model

Stage 1. A preindustrial society has high birthrates and death rates. It has a low population, and the population balance is more or less maintained.

Stage 2. Improvements in food supply and sanitation decrease the death rate, but fertility rate remains high. The population expands.

Stage 3. Postindustrial economic growth brings a lowering of birthrates, which decrease for a variety of reasons. The main one is a change in values brought on by improved economic conditions. Families "want" fewer children.

Stage 4. Both birthrates and death rates are low. The population levels out. However, low birthrates can translate into populations not capable of self-sustainability.

Stage 5. The population is still high, but it is going into decline due to an aging population.

Most of the world's poor countries are in Stage 2, with rapidly expanding populations. An example is Egypt. Its high population growth is shown in figure 3.5, where it is compared to Germany's.

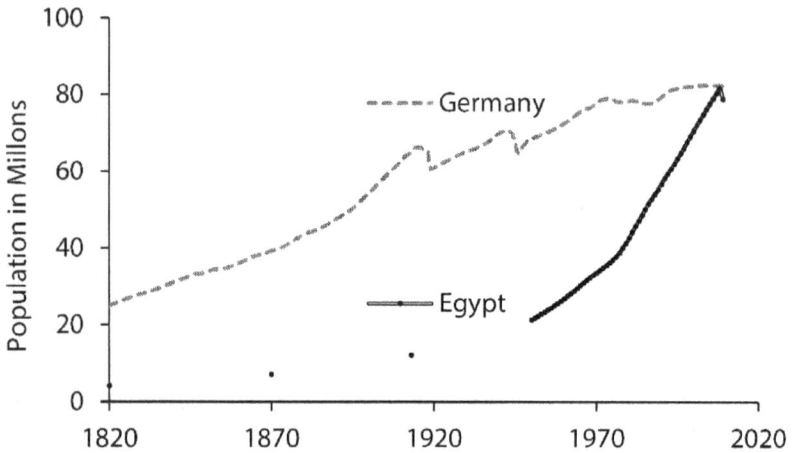

FIGURE 3.5   Population Growth of Egypt versus Germany

Egypt's Stage 2 started more than a hundred years after Germany's and, once started, took off. It is still headed north.

Germany's graph clearly shows the effects of both world wars. In the late 1970s, Germany appears to enter Stage 5, with this transition interrupted by the reunification of the two Germanys in 1990. The most recent part of the graph shows that Germany is in Stage 5, with a flattening growth curve.

A simple conclusion is that if all the poor nations make the transition to Stage 4, the curve for the world should flatten. However, an examination of fertility rates and GDP suggests that simplicity might not be the case; there could be more.

A conclusion based on figure 3.6 is that high fertility is primarily a problem of poor countries. This is true and, based on our recent discussion about the Demographic Transition, it is expected. However, there is more to the story. Of the 144 nations shown in figure 3.6, 41 have a fertility rate greater than 3.5, and of those, 35 are sub-Saharan African.

This observation raises the possibility of other issues at play. There are, but they are esoteric and not part of our central narrative. We are going to reserve that discussion for the last chapter.

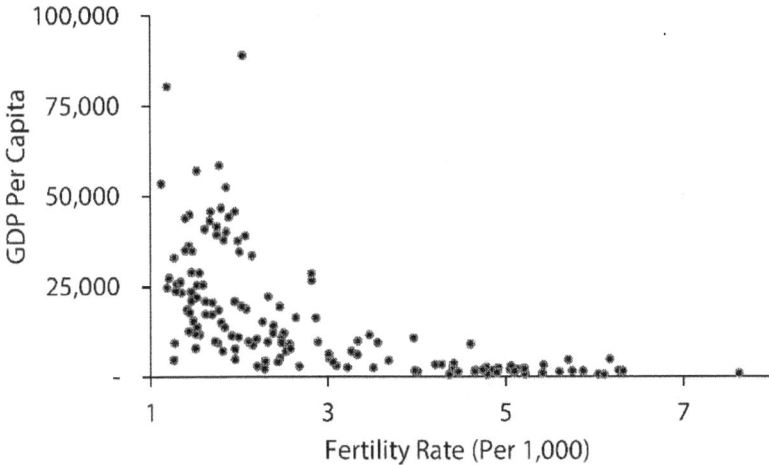

FIGURE 3.6   Fertility and GDP

Independent of cause, a consequence of the foregoing is that a poor nation with a high fertility rate reaching a Stage 4 level of the Demographic Transition might not have its population growth curve flatten out. Therefore solving the poor problem might not solve the population problem. However, we will not know until the poor nations have in fact reached the prosperous stages.

This poor–rich differential in population growth rates exacerbates the problem of economic inequality between nations. Rich nations with their flattening population curves prophesy for their citizens a high-quality future. The poor nations are not only poor; their prospects are miserable. The poor, by begetting even more, create extra hurdles to be overcome along any path exiting their low-quality existence.

We live in a world with both economic inequality and inequality of future prospects, as the growth rates of many poor nations would attest. There is a third layer to this onion: envy. Happiness is having more than your neighbor,[3] not just having a lot of stuff. Relative status is a metric for happiness. The underclass of the world will not be happy until they too can live like the wondrously affluent of California as portrayed by the marketing arm of USA Inc.: Hollywood. The next time you see a movie that is about a time and place in the United States, translate the sense of what you see and hear into the local sensibilities of a person living in a poor nation. The poor's understanding of this rich, other world is thus founded; this is where they aspire.

Solving the Poverty Problem

There is a fourth layer to the preceding metaphorical onion. This most important layer of all extends all the way to its very center. It defines the essence of our problem. It is the human face of deep, abysmal poverty. It is the children of the *favelas* in Rio de Janeiro and the mother holding her starving baby in South Sudan. It is the 47 percent of the world who live on less than $2 per day. It is you, me, our children's children, or theirs at another time and in a different place.

It would be nice to be able to write that solving the poverty problem has been uneven at best. That would be a far too optimistic statement. However, it is not for lack of trying. In the United States, President Johnson declared in 1964 a "war on poverty," and over the intervening five decades, federal, state, and local governments' accumulated spending on this war reached $16 trillion.[4] Additionally, the annual spending on welfare per person in poverty has increased approximately threefold over the past thirty years:

> Altogether, the United States has spent more than $19 trillion fighting poverty (in constant 2014 dollars). Last year alone, the federal government spent almost $700 billion, while state and local governments added nearly $300 billion more, for a total of roughly $1 trillion. That is equivalent to more than $21,000 for every person below the poverty level in America, or $63,339 for a family of three. While it is true that a significant portion of the actual money in these "anti-poverty" programs goes to families above the poverty line, the fact remains that we spend enough money on the welfare system to conceivably lift everyone who currently lives in poverty above the poverty threshold, which stood at $18,769 for a single mother with two children in 2013.[5]

If the poverty level for a family of three is $18,769, why does the United States spend $69,339 per family? It is obvious that something is terribly, terribly wrong.

While the United States was spending $16 trillion on its poor, others, primarily the West, were spending $2.3 trillion on the broader problem of global poverty.[6] Neither the efforts of the United States[7] nor those of the West[8] have significantly reduced poverty. Yes, some of this money does provide food, shelter, health care, and so on, to the poor, but the underlying, basic problem of reducing the number of poor has not been touched.

I did not arrive alone at the pessimistic conclusion I have painted here. I have company, and some are certainly more competent authorities than I am. These include William Easterly and Dambisa Moyo, published economists with experience in economic development.

Zambian-born Dambisa Moyo has worked at the World Bank and Goldman Sachs. She is currently on several boards, including the board of Barclays Bank. She has written three *New York Times* best-selling books. Her first, *Dead Aid: Why Aid Is Not Working and How There Is a Better Way for Africa,* is of particular interest to our account. In it she opines that government-to-government foreign aid has harmed Africa by fostering dependency of the population and encouraging corruption by the elite, a result exactly the opposite of what is needed to reduce poverty. She does recognize, however, the usefulness of humanitarian aid.

William Easterly, a professor of economics at New York University, is the author of several books and articles in refereed economic journals. One book, *The White Man's Burden: Why the West's Efforts to Aid the Rest Have Done So Much Harm and So Little Good,* is important to our story. His stance is similar to Moyo's, but with a difference. He divides the aid world into two factions: the planners and the searchers.

Planners are actually the topic of Moyo's *Dead Aid.* They are the ones with the large, solve-all-with-one-bold-leap-across-the-chasm approach. Planners are top-down, big-plan people who tend to impose from their superior perch. They know; they are the experts; follow their directions and the hurt will go away. Often their programs lack accountability and feedback.

The unstated, underlying assumption of the planners is that people are the same. If they were, we would not be having this discussion, and the poor would not exist as they still do after these many dollars. The operative model seems to be to show them how it is done in the first world, guide them in those ways, and the third world becomes the first. It has not happened. It will not. As we will demonstrate in this book, people are different. Expecting people to perform the same, even in the same environment, is a fool's mission.

Easterly's searchers are the bottom-up people who address narrowly defined, specific needs. They make no assumptions about people being the same or different. By working directly with the people, at the level of the individual, they do not have to make any assumptions. Actual behavior, whatever it is, is naturally baked into the solution.

The searchers represent a small-ball process, one addressing problems at the level of the individual. They do not address large, society-wide issues. The model is to incrementally solve small problems at the lowest level and the broader ones at higher levels should take care of themselves. This is the process Easterly advocates, as in his view, it has a better chance to succeed than the processes put forth by the planners.

A problem with Easterly's searchers model is that it has an underlying assumption of commonality of motive. Some populations have social institutions affecting labor for economic ends. In these societies, labor has a strong social component, one confounding what Easterly sees as primarily an economic problem.

The stunning lack of progress in solving the poverty problem, even with seemingly superfluous funds, suggests that governments and involved institutions do not understand the problem they are attempting to solve.

Two tables in Easterly's book provide insight into this issue of "not understanding the problem."[9] His book illuminates two issues, one, that aid might do harm rather than good. The second suggests that we do not know what we are doing when it comes to solving the poverty problem. Tables 3.1 and 3.2 show aid for the nations having the top and bottom best per capita growth rates over a twenty-two-year period.

TABLE 3.1   Ten Best Per Capita Growth Rates, 1980–2002

Country	Per Capita Growth 1980–2002 (%)	Aid/GDP 1980–2002 (%)	Time under IMF Programs 1980–2002 (%)
South Korea	5.9	0.03	36
China	5.6	0.38	8
Taiwan	4.5	0.00	0
Singapore	4.5	0.07	0
Thailand	3.9	0.81	30
India	3.7	0.66	19
Japan	3.6	0.00	0
Hong Kong	3.5	0.02	0
Mauritius	3.2	2.17	23
Malaysia	3.1	0.04	0
Median value	*3.8*	*0.23*	*4*

TABLE 3.2   Ten Worst Per Capita Growth Rates, 1980–2002

Country	Per Capita Growth 1980–2002 (%)	Aid/GDP 1980–2002 (%)	Time under IMF Programs 1980–2002 (%)
Nigeria	−1.6	0.59	20
Niger	−1.7	13.15	63
Togo	−1.8	1.18	72
Zambia	−1.8	19.98	53
Madagascar	−1.9	10.78	71
Côte d'Ivoire	−1.9	5.60	74
Haiti	−2.6	9.41	55
Liberia	−3.9	11.94	22
Dem. Rep. Congo	−5.0	4.69	39
Sierra Leone	−5.8	15.37	50
Median value	*−1.9*	*10.98*	*54*

The data show that those achieving success received little foreign aid while those having little or no success received substantial aid. We can deduce several conclusions, but the only two I want to raise are that

1. a poor nation can obtain economic success without foreign aid; aid is not requisite

2. receiving foreign aid could be detrimental in becoming successful

The two tables provide insight into the question, do we know what we are doing? They show performance differences between nations seemingly based on receiving or not receiving aid, issues that Easterly addresses as an economist, which he is.

Political Correctness Is Proactive Ignorance

The tables also show two different groups having different outcomes. The Confucians of East Asia dominate the ten best per capita growth rates table, either directly, as in Hong Kong, or as the dominant minority, like in Malaysia. Sub-Saharan Africans dominate the ten worst per capita growth rates table. This obvious observation suggests that differences between people could be a factor. However, Easterly never mentions this as a possible factor; therein lies the rub.

I have no idea why Easterly did not raise the issue of "different people" as possibly being a factor. It could be that he recognized that if he did, he would be expected to answer the question, why? This can be a difficult one to answer. Nevertheless, it is there staring us in our faces and should not be ignored.

Given that Easterly is an economist and concerned with economic issues, his not mentioning "different people" might be expected. Conjecture has little value in this part of our discussion, but a larger point needs making. Consequently, for the purposes of making an opening argument, I will ascribe his omission to political correctness.

I believe that by some large measure, the culture of political correctness prevents examining "people and their differences" as an explanation for poverty, or anything else, for that matter. The paradigm of "psychic unity" or equality is certainly safer. If we do bring to the discussion on poverty "people and their differences," will the thought police accuse us of blaming the victim?

Being politically correct is a commandment of social acceptance, a social norm of our modern, literate society. But it is more than just a simple social requirement. Forbidding discourse of certain content is more than pigheaded. It is potentially responsible for killing mothers, babies, small children, and other innocents, because solutions to important societal problems based on non–politically correct, but nevertheless factually accurate, ideas are not allowed.

I have no special claim to knowledge of "people and their differences" as a factor of poverty. If it is, and I believe that it is, then letting political correctness prevent understanding of the problem means not solving it.

We might disagree on several things, but refusing to bring an idea to the debate because it is a taboo thought is censorship, a type with potentially serious, even fatal consequences. Political correctness, as currently exercised by the progressive left, especially those in academia and the media, is a form of censorship, of intolerance of certain ideas. Censorship, especially of ideas within learned debates, should have no standing in our free society, and we must not tolerate such intolerance.

The philosopher Karl Popper brings two insightful quotations to the conclusion of this chapter:

> If we extend unlimited tolerance even to those who are intolerant, if we are not prepared to defend a tolerant society against the onslaught of the intolerant, then the tolerant will be destroyed, and tolerance with them.[10]

> True ignorance is not the absence of knowledge, but the refusal to acquire it.[11]

This is where the politically correct reside. They are not only truly ignorant; they are proactively so.

CHAPTER 4

Inequalities' Measure

Where do we start to look for the reason nations have inequality in their economic outcomes? Often, the approach to such queries is with race or ethnicity. For example, some will try to explain why this or that group is successful or not by invoking a stereotypical character of their race or ethnicity. The cause is simple, race and the effect expected. It is the use of stereotypes to explain, an approach used by the analytically challenged. No actual analysis needed, just go to the lookup table ascribed to the race.

Even though stereotypes can be correct, they are far too simple and, therefore, potentially misleading. Besides, they offer no explanation as to why; they are a sort of "just because." We can do better, and we will. However, before beginning, we should consider how race and ethnicity apply to our question.

Race and Ethnicity

We, or more precisely anatomically modern *Homo sapiens,* originated in Africa, but the sequence and timing of our departure from humanity's homeland to the rest of the world is in debate. For example, did we migrate in one or two distinct waves, with one much earlier? Adding confusion to our genetic heritage is the fact that, once free of the African continent, some, but not all, bred with earlier escapees, the Neanderthals and Denisovans. What seems to be generally accepted is that our migration probably started no later than fifty thousand years ago, and that is the date we will use.

Evolution is all about adapting to changing environments, and from our first step out of Africa to the Industrial Revolution, it has experienced two major changes. First, by leaving Africa, we changed *where* we made a living, and then, after about forty thousand years, or ten thousand years ago, and with the Neolithic Revolution, we changed *how* we made a living.

When we migrated away from warm, sunny Africa and toward the colder, less sunny North, we had to make physiological adaptations, and when we decided that living high in the Andes or Tibetan Plateau was a good idea, we had to make further adaptations. Actually, we did not do anything; evolution did. If it had not resulted in our adapting to our new environments, we would still be stuck in sunny, warm climates at a reasonable altitude, and we would all have dark skin. By changing the location of where we made a living, we became lighter skinned, our hair changed, and we made adaptations to the cold. Furthermore, moving to where mosquitos could not live, we lost some adaptations pertinent to our old homestead, including defenses against malaria.

Changing where we lived might have induced some groups to undergo unspecified behavioral adaptations for unspecified reasons. However, because behavior does not leave footprints, as do bones, we do not know. The main point is that these changes in our physiology mark our starting point for forming races, but lacking reason for behavioral change, we probably maintained our unity of psyche. This means that forming races resulted in different physiologies, but lacking a reason for change, psychologies probably remained similar.

After we had been away from our original homeland for about forty thousand years, and at the start of the Neolithic Revolution, we had races, each with different physiologies. However, the other half of the coin, psychological change, likewise requires reason, and that was not to occur until we changed how we made a living. This happened with the start of agriculture and the Neolithic Revolution, the second big change in our history.

If you enjoy fresh-picked corn on the cob, fried green tomatoes with shrimp remoulade, or simply some steamed broccoli with a tad bit of mayonnaise, thank global warming and climate change. This is not the one currently in progress but the one at the end of the last ice age, the end of the Pleistocene and the start of the Holocene. Back when, before we became farmers with the Neolithic Revolution, it was just too cold and dry for plants; farming was not a viable option. Starting about ten thousand years ago, global warming enabled our transition to making a living with agriculture.[1]

When we changed from nomadic hunter-gathers to sedentary farmers is when we expect extensive psychological changes to have occurred. This transition was a huge change in our evolutionary history because it forced all who transitioned to change behavior. For our entire prior evolutionary history, we had been nomadic foragers practicing hunting and gathering. Now, we could settle down, and not having to move every few days, we could acquire stuff like storage jars and a permanent fire for cooking. Being stationary, we could have houses, and therefore we could literally form households. To succeed at our new way of making a living, we had to become smarter; domesticating plants and animals for our needs is a sophisticated task. Because we produced our own sustenance and were no longer subject to the whim of nature, we could form larger clans. Forming larger clans is adaptive, because when in conflict with other groups, whether in offense or defense, bigger is better. We could fill pages describing the entire behavior-changing requirements wrought by our new way of living, but we have made the point.

The foregoing point about the Neolithic Revolution initiating psychological changes does not rest on hand waving; there is indirect evidence. In *The 10,000 Year Explosion: How Civilization Accelerated Human Evolution*, Cochran and Harpending make the case that human evolution is recent and copious. This is also the case made by Wade in *A Troublesome Inheritance: Genes, Race, and Human History*. Much of the evidence is in the history of genetic changes that geneticists are now able to decipher. The evidence shows the frequency of genetic change having increased since the start of the Neolithic Revolution.

Given the foregoing narrative, this result is expected. With the change in where we lived, we would expect some increase in genetic changes reflecting physiological adaptations. On top of this first change in physiology is the larger one caused by the later Neolithic Revolution, one requiring multiple psychological changes. This is when we start to lose our unity of psyche.

When we consider differences between current populations, we expect them to be a consequence of both the change in where we make a living and how. Even though we expect both transitions to be influential, significant behavioral changes probably were the result of the change in how, not in where, we made a living. Consequently, because race is primarily a consequence of where and not of how, it is a poor candidate to explain the psychological differences between people. If this model is correct, multiple psychological changes probably did not occur until after race was established, and consequently, race should have little or no influence on our psychology. If we are interested in differences in behavior, and we are, we will have to use something other than race as a means of classification.

If all areas of the world changed at the same rate, ethnicity would be useful. We can see why by comparing the history of China to that of sub-Saharan Africa. In China some four thousand years ago, during the Xia Dynasty, there were approximately three thousand polities, and by the Qin Dynasty, in 221 BCE, these had reduced in number to one.[2] Sub-Saharan Africa with its hundreds of tribes has made no such consolidation. How could we ever compare the now single ethnicity of Chinese to the multitude in sub-Saharan Africa? We cannot.

We can certainly compare one ethnic group in Africa to another and arrive at valid conclusions. However, for our purposes, we need to compare sub-Saharan Africa as a complete entity to other compete entities, whether they are unified or not. Even though the truths we need for our analysis probably reside in "ethnicity," as a basis for comparison on a global scale, ethnicity is unworkable.

Preparing for the Analysis

When first starting to address the question of economic inequality, there was no preconceived notion of where to start. Given our topic, making an economic assay of the world as it exists today seemed like a logical starting point. I suspected that the assay would show the nations of the West being at the top of the economic heap and those of sub-Saharan Africa at the bottom. I had no real feeling of where countries like Russia or Tunisia or Thailand would fit into the scheme of things. Most importantly, I had no clue even if there was a scheme.

To start, I got some numbers, data that I could examine and perhaps be able to detect some order in—a scheme, if you will. The data selected are from the World Bank[3] and include 249 nations. The data comprise GDP per person, corrected for purchasing power parity (PPP),[4] a measure of the economic outcomes of nations. We will use this data set throughout the book as our primary metric.

As expected, once the data were rank-ordered from rich to poor, the nations of the West were at the top and sub-Saharan African nations were at the bottom. However, a more in-depth examination showed that countries with the same religion had similar economic outcomes. For example, the Muslim nations approximately fell into a group. Schemes with apparent rationality are nice to find on any journey of discovery.

On initial examination, there is no reason to consider that one's God or how a nation prays has any bearing on economic outcomes. However, there is more to religion than spirituality. Religion provides the rules of behavior for societies and their individuals. Perhaps there is a rational association between religion and economic prosperity.

The initial observation that economic prosperity has an association with religion suggests that religion should be our organizing principle for analyzing data. It will be useful, but there are other considerations. We plan to use Darwinian Theory to derive understanding, and that creates a requirement for "common evolutionary history." We will find that, generally, people belonging to a

TABLE 4.1   Data Bins

Religion / Geography	Comments
Western Christianity	This is the Church of Rome, including both the Protestant and Catholic branches. These are the nations of Western Europe plus their offspring of Canada, the United States, New Zealand, and Australia. We will call this Western or the West.
Eastern Christianity	This is the (Eastern) Orthodox religion. It comprises primarily Russia along with its Slavic cohorts and some Balkan nations.
Confucian	These are the nations of East Asia, all having the common denominator of the Confucian philosophy. We will call this Confucian or the East.
Islam	These are the Islamic nations from Morocco in the west to Pakistan in the east. Indonesia and Malaysia are included in Southeast Asia.
Hindu	This is India.
Latin America	Latin America is not part of the West. Even though it is Catholic, its heritage does not include that of the Roman Empire and the early Church. It is unique.
Southeast Asia	Many of these nations are Buddhist. In their early periods, all received strong Hindu and Buddhist influence from India.
Sub-Saharan Africa	This is not associated with any major religion.

common religion meet this requirement. Another aspect of "common evolutionary history" is adjacent geography, and that too is a property of most major religions. Some religions are associated with a given civilization, making it a possible organizing principle, and then we could simply use geography.

None of this is perfect. Where appropriate, we will use religion, and where not, we will use geography. Table 4.1 lists the eight bins we will use.

Western Christianity consists of the nations of Western Europe along with their offspring in Canada, the United States, New Zealand, and Australia. It has two religious subgroupings: the original Catholics and a large splinter group, the Protestants. When we do our analysis at the end of our story, we will separate the two, but for our main journey, we will treat them as one. The Catholic nations include four of the five PIIGS (Portugal, Ireland, Italy, and Spain), France, Poland, Hungary, and seven others. The Protestants are the former British colonies of New Zealand, Australia, the United States, and Canada. They include the Nordic countries (Finland, Sweden, Denmark, and Norway), the United Kingdom, Germany, Switzerland, and four others. Western Christianity contributes 13 percent of the world's population.

Eastern Christianity, or the Eastern Orthodox religion, with only 4 percent of the world population, is our smallest. It consists of Georgians, Greeks, and Romanians, along with the eastern part of the Slavic ethnic group. The Slavic countries in this group are Belarus, Bulgaria, Macedonia, Moldova, Russia, Serbia, and Ukraine. The Slavs in the western part, such as Poland, are Catholic, making the Slavs a case of one ethnic group falling into two different religions.

Confucianism is not actually a religion; it is a philosophical and ethical system based on the writings of Confucius. The Confucian ethic itself is important in understanding economic outcomes, and it makes sense to include the entirety of the Confucian ethic under a single umbrella. Our Confucian nations include Japan, Vietnam, Korea, and China. They also include the Han Chinese–dominated nations and nation-states of Singapore, Hong Kong, and Taiwan. This group accounts for 24 percent of the world's population.

There are twenty-nine Islamic nations in our analysis. They exist in a swath of land that starts with Morocco on the Atlantic and runs eastward to Bangladesh. The last two on that traverse to the Pacific Ocean, Indonesia and Malaysia, have more in common with their neighbors, and we will group them with Southeast Asia. Our grouping of Islamic nations accounts for 13 percent of the world's population. Within this group is a variety of ethnic groups, including Turks, Arabs, Berbers, Bengalis, and Persians. The Islamic nations include Arab countries like Saudi Arabia and Egypt. Other populous Islamic nations are Bangladesh, Turkey, and Pakistan.

The Hindu religion consists of India and Nepal and is 19 percent of the world's population. Approximately 80 percent of Indians are Hindu, a Dharmic religion, and 13 percent are Muslim. This makes India the second most populous Islamic nation after Indonesia. Other Dharmic religions in India are Buddhism, Jainism, and Sikhism. India is also home to the economically important and "dominant minority" Parsee, Zoroastrians, originally from Persia. It has 645 recognized indigenous tribes.

Some might consider Latin America as belonging to the Catholic religion. Because of their conquerors, they are Catholic, but they are not part of the Catholic heritage that began with the Roman Empire. What we see now is the result of two processes. Approximately 80 to 95 percent of the indigenous population died from diseases imported by their conquerors,[5] and in the following years, they mated with the Spanish and Portuguese to create mestizos. Adding to the pot, particularly in Brazil, are the people of African descent. Although there are pure Europeans and indigenous people in Latin America, mestizos are the majority. Latin Americans are a new people and do not belong with any older, established group.

With 9 percent of the world population, there are eighteen countries making up Latin America. Brazil is the only Portuguese colony; all the others were Spanish. Besides Brazil, the big players are Argentina, Chile, Mexico, Venezuela, Colombia, and Peru. All six countries of Central America are relatively small.

We could treat Southeast Asia consistently with our organizing principle and create a bin for the Buddhist religion. A Buddhist bin would include Myanmar, Thailand, Laos, and Cambodia, which makes sense. However, it would also include Mongolia, which does not. In addition, the Philippines, with its Catholic population, would logically have nowhere to go and, consequently, would not be included in our analysis. A major point is that the nations of Southeast Asia have a degree of common history. They all have had cultural input early from India in the form of Hinduism and Buddhism. Although Indonesia and Malaysia are currently Islamic, in their early years, they were much the same as the others, tied together by Indian influence.

The foregoing discussion suggests a geographical approach. We could treat Southeast Asia as Southeast Asia and not depend on religious labels. This would remove Mongolia from the list and bring the Philippines into the conversation, both making sense. The normal grouping of Myanmar, Thailand, Laos, and Cambodia would remain. The big change would be moving Indonesia and Malaysia from the Islamic bin and placing them in the one for Southeast Asia. This removes the most populous Islamic nation, Indonesia, from the Islamic bin. However, separating these two from their Ottoman Empire coreligionists seems logical. This is not perfect, but it is what we will do. We will have a Southeast Asian bin and not a Buddhist bin. These nations comprise 7 percent of the world's population.

Much of Africa north of the Saharan Desert is included in the Islamic bin. However, the main part of sub-Saharan Africa does not belong to a single, unifying religion. It only established an association with major religions after discovery in the fifteenth century. We will use a sub-Saharan African region. It has 12 percent of the world's population and consists of forty nations. The top ten most populous nations are South Africa, Nigeria, Tanzania, Kenya, Uganda, Ghana, Côte d'Ivoire, Mozambique, Madagascar, and Cameroon.

In addition to Buddhism, we will not use two other religions as part of our organizing scheme. One is Oriental Orthodox Christianity and consists of Albania, Ethiopia, and Eritrea. They comprise only 1 percent of the world's population. Even though Ashkenazi Jews are part of our story, we will not have a Judaism bin. It has relatively few adherents, and Israel, with only 0.1 percent of the world's population, is not included.

Organizing the Data

With an organizing principle of religion and geography, we can organize the data. However, we first need to do some trimming. Populations in the 249-nation data set of the World Bank range from 1,357,380,000 for China to 9,876 for Tuvalu. To minimize distortions in the analysis, we need to exclude the smaller nations so they will not have an effect disproportionate to their size. To this end, we will exclude nations with a population of less than 1 million. There are fifty-six (22 percent) nations in this category. The excluded nations include Fiji, Iceland, Belize, and Grenada.

The cutoff at a population level of 1 million is not arbitrary. Cutting off at one hundred thousand eliminates only twenty-one (8 percent) nations. There are 127 (51 percent) nations having a population less than 10 million, and eliminating them reduces the data for analysis by more than 50 percent. A Goldilocks value of 1 million is just about right.

This book is in part about wealth, a concept that George Gilder[6] in his best seller *Wealth and Poverty* divided into two types. One, like oil in the ground, is a "consequence" and without "cause." The other is a "cause," represented by the relatively barren islands of Japan and Great Britain. It is the "cause" of wealth, people, which we are concerned with here, not some fortuitous "consequence" of the Carboniferous period.

The World Bank data set includes total natural resources rents percentage of GDP in 2013. I used these data to adjust the GDP value for the effect of oil, forest, minerals, and other fortuitous resources. What remains should be what the people make, the wealth of "cause."

Not all nations in the data set have values for GDP. They are Myanmar, North Korea, Somalia, the West Bank and Gaza, and Syrian Arab Republic. They are not part of our analysis. Neither Argentina nor Taiwan is in the data set, but substitute values exist in a similar data set from the International Monetary Fund.

Ten remaining nations do not fit into any of the groups and consequently are not included. We have discussed Israel, Armenia, Ethiopia, and Eritrea. The former French colony Haiti and the former British colonies Jamaica and Trinidad and Tobago do not belong with their Latin American neighbors. Papua New Guinea, without any sort of classical cultural identity, is a real outlier. Lacking an appropriate association, Buddhist Mongolia has nowhere to go. After organizing and clipping, 145 countries made the cut.

We now have eight religions or geographical areas into which we will bin the data of 145 nations. The results, "Wealth of Nations," are in figure 4.1.[7]

Sub-Saharan
Hindu
Muslim
SE Asian
Latin American
Confucian
Orthodox
Confucian (Less China and Vietnam)
Western

| - | 10,000 | 20,000 | 30,000 | 40,000 |

GDP Per Capita

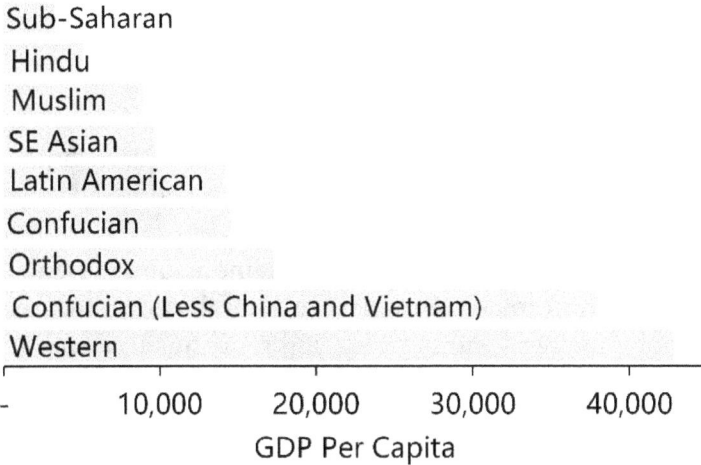

FIGURE 4.1   Wealth of Nations

There are two entries for the Confucians. One includes its entirety, whereas the other excludes China and Vietnam. In a subsequent chapter, we are going to discuss extractive versus inclusive
governments, and we will see that this concept acts as a gateway to
economic success. China and Vietnam have extractive, communist governments, limiting the native ability of their populations to
succeed. This difference provides rationality for separation.

Another point is that China's extreme size masks the reality
of the other Confucian nations. These nations are presently economic successes, and that is going to be an important point in our
analysis.

By separating China and Vietnam from their brethren in the
East, are we fiddling with the data? Yes, but as will be evident by
the end, it makes sense and will sharpen our accounting. The East
has not yet reached the West economically, but when China finally
completes its economic ascendency, it will. It is time the West un-
derstands that fact.

For purposes of our discussion, we will assume that if both
China and Vietnam had liberal democracies like their Confucian
brethren, they too would exhibit economic success. Continuing
with our make-believe world assumption, the bar labeled "Confu-

cian (less China & Vietnam)" represents the *potential* of the Confucian nations, admittedly making the Confucian world richer than reality.

With this assumption, the world as expressed in figure 4.1 has two distinct groups. There are the Rich West (Protestants and Catholics) and East (Confucians); there are the Middle-Class Latin Americans and Orthodox Christians. Then there are the Poor, consisting of the Southeast Asians, Muslims, Hindus, and sub-Saharan Africans. Contrary to conventional wisdom, the economic world is not simply "the West and the Rest." It is more nuanced and, as we will learn, much more interesting.

New Questions Raised

Our original question of why some nations are rich while others remain poor just became more complex. Answering it remains our primary goal, but more questions now exist.

Biologically, modern humans arose in Africa many thousands of years ago and over time migrated to all parts of the world. From this perspective of Africans as the source of humanity, we perhaps can consider "Africa" to be the baseline human condition. A question that arises is, why didn't Africa rise economically with the rest? Does the same reason apply to the other three nonperformers, the Hindus, Southeast Asians, and Muslims? If not, then what is the explanation for their lack of economic performance? Hopping to the other, wealthier side, why did the West become so successful? Did Confucians get to their high level by the same process as the West, and if not, what are the differences? Last, how do we explain the Middle Class of Latin Americans and Orthodox Christians? Why did the Rich become rich, and why do the Poor remain poor? Why are we the way we are?

CHAPTER 5

Access to Women

◇◇

Evolution's Handmaiden

We noted in the preface that the seed for this book was Gregory Clark's *A Farewell to Alms: A Brief Economic History of the World.* His book earns status as seed for our sociobiological narrative because he brought into his economic analysis an important Darwinian observation. Using data from testators' wills, he observed that the rich English of preindustrial times had more surviving progeny than the poor.

This is what Darwinian Theory is all about—differential survival and reproduction; it does not get more basic than that. What better foundation to develop an explanation of why some are rich and others remain poor than an observation that the rich are evolutionary fitter than the poor? There is, of course, more, a lot more, and we are going to use most of the remainder of our book to fill in the details.

With this chapter, we start the development of a model that, in the end, will explain why some nations are rich while others remain poor. In building this model, we are not going to take the usual top-down, deductive approach. Rather, using first principles and synthetizing the general from the specific, we will follow a bottom-up, inductive path.

Darwinian Fitness

Men compete for a variety of reasons, and access to women is at the top of the list. Getting your genes down the road is far more important than the impromptu footrace or wrestling match. Throughout our story, we will see that competition between men for sexual access has been a major driver in the evolution of human nature. It has been evolution's handmaiden.

Differential reproductive success or fitness is a primary element in the Darwinian evolutionary process. Individuals with high fitness leave behind for future generations more copies of their genes along with the heritable portions of their culture than those less fit. The important consequence of this process is that, over generations, the traits of populations shift to reflect those of the fit individuals. Nature makes copies of the successful, not the failures. Accordingly, populations change and become different.

The starting point of our model is two sets of observations concerning differential reproductive success. The first, which we have introduced, is about the rich of preindustrial England. The second, located at an earlier point in the spectrum of human development, is about a primitive, pre-state tribe in the Amazon, the Yanomamö. It is with these examples that we can start to develop our model of the evolution of human nature.

In his studies of the Amazonian Yanomamö tribe, anthropologist Napoleon Chagnon observed that *unokais,* warriors who have killed in war, had more surviving offspring[1] than non-*unokais*. In a similar fashion, but by using seventeenth-century testator data from England, the economic historian Gregory Clark observed that rich men had more surviving offspring than the poor did.[2] The common element in these two observations is that of differential reproductive success or evolutionary fitness. In their respective societies, the *unokai* and rich Englishmen are fitter than others are.

Darwinian Theory predicts that heritable traits playing a role in reproductive and survival success will increase in their respective populations. These traits include both alleles (gene variants) and the heritable portions of applicable culture. This results in the

English population becoming better at achieving economic success and the Yanomamö becoming fiercer warriors.

We start the development of our model with these two actors. One we can refer to as *Homo economicus,* which has as an archetype a successful English merchant. The other, *Homo bellicus,* has as a model exemplar the *unokai* warrior. Each is symptomatic of a higher-order social condition, the modern economic society and the tribal warrior society, respectively.

Status and Access to Women

One of the proposed titles for an earlier version of this book was *Don't Ask Such Stupid Questions! Women! Women! Women!* Among other issues, it was excessively long. However, its *raison d'être* is that it contains the kernel to answering the question, why are people different?

The proposed title came from *Noble Savages: My Life among Two Dangerous Tribes—the Yanomamö and the Anthropologists*:

> My many hours of taped interviews of old Yanomamö informants who told me about their immediate and distant past and of the many wars they fought before they moved to the villages in which I found them are very clear about the importance of women in their conflicts. One of my old informants, Dedeheiwä, a renowned shaman from Mishimishimaböwei-teri, is typical. After days and days of interviews during which I kept asking him about the cause of the fight at every place where a village split into several hostile villages, he said in exasperation: "Don't ask stupid questions! Women! Women! Women! Women! Women! They screwed all the time and made a noise like wha! wha! wha! when they screwed. Women!"[3]

Does this mean that we are going to accuse "women!" of making people different? No, we are going to put the blame on men. They are the proximate cause because their genes drive them to seek access. Nevertheless, even though they have no say in this matter, women, or rather their biology, remain as the ultimate cause, as they control the limiting factor of reproduction—eggs. A woman can have only so many children, but a man has no such biological limit.

In the civilized, modern world, men do not have to kill in war to gain access to women. There are means for gaining access other than being a *unokai*. Former national security advisor and secretary of state Henry Kissinger reportedly said, "Power is the ultimate aphrodisiac."[4]

The primitive Dedeheiwä and the sophisticate Kissinger were saying the same thing—high status confers access to women. In Dedeheiwä's case, he is saying that access to women comes with being a superior warrior, because being a superior warrior confers status. Kissinger is saying that it is power, or its functional equivalent, money, that provides access to women.

In both instances, biology is the ultimate driver. Men have the natural, evolved drive to maximize the number of copies of their genes in future generations. Correspondingly, a primary metric of evolutionary success is having more surviving offspring than others do. This is Darwinian reproductive fitness.

The Yanomamö

Chagnon spent twenty-five years studying the Yanomamö, a primitive tribe living in the Amazonian rain forest of southern Venezuela and north central Brazil. Over those years, he spent in the field a total of sixty months over twenty separate trips.

The Yanomamö consist of some thirty-five thousand people living in 200–250 villages. The population of the villages is between fifty and two hundred persons. They occupy an area of approximately 120,000 square kilometers, resulting in a population density of one person per six square kilometers.

They have a subsistence economy. They make their living by practicing slash-and-burn farming and are knowledgeable horticulturists. They also engage in hunting and gathering.

All the people of a village live in a large oval building called a *shabono*. It has internal partitions that form "houses" for the family units. Each *shabono* has a garden. Men do the initial work of land clearing, and the women do the actual gardening. They grow plantain, tobacco, manioc (cassava), ohina (a taro-like tuber), sweet potato, and hot peppers. They cultivate fruit trees such as avocados and papaya. Nonfood crops include cotton for hammocks.

Approximately 80 to 90 percent of their food comes from their gardens. This makes the women the primary producers and establishes them as important economic assets. Only men participate in hunting, and this activity is where they spend about half their time. They hunt wild pigs, monkeys, tapirs, armadillos, caiman, birds, and rodents. They also gather caterpillars, grubs, palm fruit, honey, and a variety of small critters. They fish when the fish are abundant and easily taken.

Both men and women work about four hours per day, spending the remainder in leisure activities. In the afternoon, women socialize and the men usually snort hallucinogenic drugs. Given this work schedule, they apparently have labor bandwidth well in excess of their "needs."

Given the foregoing, they do not appear to be resource limited, at least on a short-term basis. However, over a few years' time, there is a reduction in both available wild game and soil nutrients of the garden. This resource depletion forces the village to relocate every few years.

Two factors determine the village size. At the high end, it is the amount of accessible wild food, especially animals like pigs and tapirs. At the low end, vulnerability to attack from other villages dictates a minimum size, as small villages are more vulnerable to raiders than larger ones.

A village can fissure or split up into two smaller groups. This can be simply the consequence of two groups not getting along or simply a village becoming too large. However, fights over women are often the cause. Villages will get to a population of 100–150 persons before fissuring, but some get larger.[5]

Yanomamö villages are patrilineal, as they emphasize descent through the male line. Per Richard Dawkins's Selfish Gene Theory,[6] the men of such a village would have their genetic interests aligned, resulting in a high level of cooperation. In war, men having a higher level of cooperation should dominate over those not as cooperative. Additionally, a high level of within-group cooperation should help reduce internal conflicts, allowing villages to grow to larger sizes before fissioning.

The preceding description appears to be in discordance with Malthusian Theory, which has population growth controlled by growth in food production. With the Yanomamö, food is more than adequate, and they could easily reproduce more. Yet, there is no evidence of population growth. One explanation is that they limit population with abortion and infanticide. Another is their proscriptive rules on having sex. Moreover, there is of course population control via their chronic state of war. Their self-referential term is "fierce people,"[7] establishing war as an important aspect of their culture.

The Yanomamö and Us

The Yanomamö, like many primitive societies, have no written language; their numbers are simply 1, 2, and more than 2. We have no idea what they were like at the dawn of the twentieth century, let alone the first. Are they a historical representative of the people who were to become our modern, economically developed societies? Is their behavior the behavior of your or my ancestors sometime before the dawn of almost anything? We probably will never know.

Is it important? Yes. If we were then as they are now, we have undergone major changes, especially in behavior. Knowing "how we were" tells us something about "who we are." More importantly, understanding the processes resulting in the changes making us "us" will tell us "why" we are, a more insightful question than "who."

In their book *Demonic Males, Apes, and the Origins of Human Violence*, Wrangham and Peterson develop a thesis that men act violently and make war because of traits inherited from our chimpanzee-like ancestors.[8] Depending on whom you ask, humans and chimpanzees have either 99 percent or 94 percent identical DNA. Given this close coupling, their thesis is almost expected.

A *unokai*, you, and I all have the same chimpanzee-like ancestor. All three of us arrived where we are today on some continuum of development, not via discontinuities. Our paths might have been uneven and different, but all were continuous. A relevant quotation comes from Fukuyama:

> For the account of the state of nature given by Hobbes, Locke, or Rousseau to be correct, we would have to postulate that in the course of evolving into modern humans, our ape ancestors somehow momentarily lost their social behaviors and emotions, and then evolved them a second time at a somewhat later stage in development.[9]

We are going to assume that the Yanomamö, or something like them, represent our behavioral ancestors on some continuum of developmental history.

Our modern *Homo economicus,* where rich men get preferential access to women, was probably a Yanomamö-like *Homo bellicus* at the start of the Neolithic period, about 10,000 BCE. Between the "warrior culture" of the Yanomamö and the societies of modern nations with market economies exists a chasm, an economic, cultural, and social differential. We were all the same at some distant time in the past, but now we are not. We need to understand how we became the way we are, but more importantly, we need to understand the why of the paradigm shift that occurred. We will do both in the following chapters.

War and the Yanomamö

Figure 5.1, "Percentage of Male Deaths due to Warfare," is derived from data taken from *War before Civilization: The Myth of the Peaceful Savage.*[10] It shows that approximately 37 percent of Yanomamö Shamatari male deaths are attributable to war, while another Yanomamö tribe, the Namowei, had a corresponding 24 percent. These high rates suggest that war is probably a major factor in population control for the Yanomamö as well as the other noted tribes. However, the main point is that war for the Yanomamö is an important cultural force.

There is another point, one that will come into play later. Pre-state tribes engaged in war on a frequent basis. This makes security an issue for survival, and the evolution of human organizations

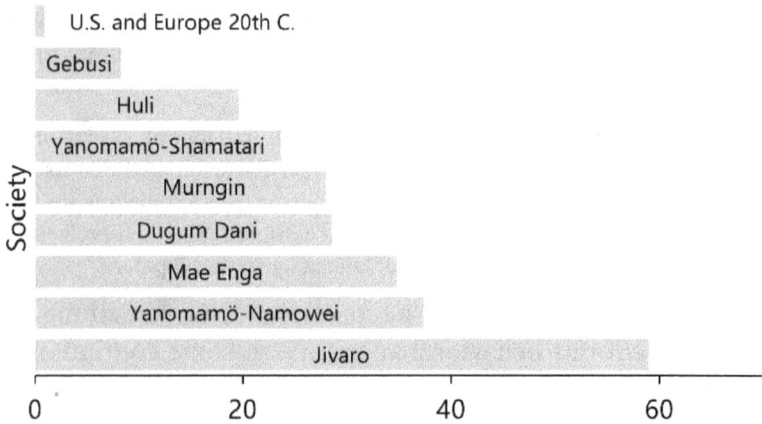

FIGURE 5.1   Percentage of Male Deaths due to Warfare

should reflect that truth. Humans who formed organizations best offering security should be the ones surviving.

With the Yanomamö, war and its preparation are major preoccupations. We can get a general explanation for war based on evolution from Low, who states "that lethal conflict exists because individuals and families have profited from assuming the risks of lethal conflict under specific conditions, over evolutionary time."[11]

Given the proper set of conditions, war provides an evolutionary payoff. For example, with the Aztec, commoners could advance by capturing enemy warriors in battle, and conversely, humiliation was the reward for men who never captured an enemy warrior.[12] Moreover, like for the Yanomamö, being a successful Aztec warrior was the ideal.[13]

With the Yanomamö, ritual war activities scale from low level— it will hurt, but you will not die—"side slapping" to potentially fatal club fighting. In addition, there is actual war itself. War also scales. At one end of the scale, there are quick forays to kill one or a few "other guys" for the purpose of retribution. Along the way, expect kidnapping of some women. At the other end of the scale are infrequent but very deadly large-scale events. An example would be a raid wiping out much of an entire village, killing all the men and kidnapping the women and children.

Part of this culture of war is the Yanomamö penchant for violence that is not directly related to war. For example, children are encouraged to slap an adult if they are unhappy with the adult's actions. From day one, boys are encouraged to be aggressive and brave—traits any warrior society would value. Yanomamö men take a great deal of pride in being fierce warriors. This cultural expectation is similar to that of the Aztec, where "the main purpose in life was to be a soldier."[14]

War in pre-state societies, as among the Yanomamö, although not universal, was highly common. Keeley reports on three cross-cultural surveys relating to the frequency of war, and his conclusion was that the vast majority of these societies were involved with war.[15] One table has 85.7 percent of Yanomamö-like societies in continuous war, 14.3 percent in frequent war, and none never at war.[16] Given Keeley's data, the Yanomamö chronic state of war, as reported by Chagnon,[17] is expected.

War and Access to Women

A primary reason for war among the Yanomamö and other, similar tribes is access to women, not access to resources such as land, water rights, or food.[18] However, wars between Yanomamö villages were not always overtly about getting women. Revenge for a killing in a prior war and disrespect were also common causes. However, in the larger scheme of things, Yanomamö wars were mainly over access to women.[19]

Both Keeley[20] and Chagnon[21] use a massacre in about 1325 at a site in South Dakota (the "Crow Creek site") as an example of war for women. The Crow Creek people lived in a village consisting of several earth lodges (packed earth and pole support) and, like the Yanomamö, were dependent on cultivated foods. Raiders attacked and killed most of the residents, leaving the remains of approximately five hundred people. The remains showed signs of mutilation, such as the removal of scalps and beheading. Absent were the remains of young women (twelve to nineteen years of age) and young children. The belief is that they became captives and extra mates of the raiders. Access to women as a motive for war is common in societies similar to the Yanomamö.[22]

The Yanomamö practice both polygyny and infanticide, where female infants are selectively killed over males. These cultural practices can result in a situation where there may not be enough women available as mates and a significant percentage of the young male population might be without.

Given this situation, a man wanting a mate can take one of two actions. He can take one from another man in his village, or he can take one from another village. As most of the men in his village are his blood relatives, it makes sense for him not to attempt a local sexual coup. A reasonable course of action is to form an association, a marauding band of young males, for example, and go steal a few. Of course, the act of stealing women can easily result in harm or even death to some of the defending males. In turn, this second group would want retribution and, while they were at it, to take a couple or three of the original initiators' women. Get those iterative processes going, and it might never end; it has not.

In the foregoing scenario, war that results in male deaths is a process that should reduce future war, as it decreases the aggregate demand for access. A reduction in the male population translates to a lessening demand for access to women. A balance between availability of female mates and male war deaths could be an equilibrium condition affecting frequency of war.

In addition to access to women, there is an economic aspect to the Yanomamö raiding another village for women. As the women, with their gardens, account for about 80 to 90 percent of the food produced, adding a woman to a polygynous Yanomamö family increases that family's net food production. In a similar manner, children are also economically valuable in that after a certain age, they assist the women in the garden. From an economist's perspective, both women and children in the Yanomamö society are assets of production.

Status Increases Fitness

Chagnon spent much of his efforts in the study of the Yanomamö in acquiring genealogical data, and using these data, he was able to determine a trait leading to improved evolutionary fitness. He found that *unokais,* men who have killed in battle, had on average three times as many offspring as non-*unokais,* a significant reproductive advantage.

Unokai is a significant title, and its reward is part of an important ceremonial ritual. It is analogous to a modern-day warrior receiving a military commendation such as the Silver Star. A *unokai* is a recognized warrior of high status, not a deranged killer. However, the important consequence of *unokai* status is that the Yanomamö culture, in lieu of passing out Silver Stars, rewards its successful warriors with more wives. "For all ages pooled, *unokais* have an average of 1.63 wives compared to an average of 0.63 for non-*unokais.*"[23] Genealogical data show clearly that achieving *unokai* status confers Darwinian fitness. As Chagnon states, "*unokais* have, on average, 4.91 children compared to same-age non-*unokais,* who average only 1.59 offspring each, that is, *unokais* have three times as many offspring as non-*unokais.*"[24]

Among the Yanomamö, is the status of *unokai* the only basis for having extra wives? No, the headmen of Yanomamö villages also have extra wives. However, this is a distinction without a significant difference, as the headmen are also usually *unokais.*

Evolution is about differential reproductive success, and it is about the resultant change expressed in populations. Bobbi Low, in discussing war over access to women as an agent of such change, stated, "In preindustrial societies like the Yanomamö and the Meru, a man's lifetime reproductive success was likely to be closely correlated with his performance in war."[25]

Men in tribes similar to the Yanomamö and Meru, who are successful in war, will have their heritable traits, both genetic and cultural, differentially survive over those of men not similarly successful. The frequency of their genetic alleles will increase in their respective populations, which, over time, will reflect the genetic dominance of successful warriors. The same can be said of the heritable cultural traits.

There are a couple of ways of looking at the issue of "war giving men access to women." The obvious, represented by the aforementioned Crow Creek site, is direct and unambiguous—attack a village, kill off the men, and take the women and children. Overlooked in this scenario is the question of who gets to mate with the abducted women. This is the bottom-line question. Only the actual recipients of these spoils will have increased reproductive success, not the entirety of the raiding party.

The answer to this question may not be important in the case of the Yanomamö and other similar societies. We know from Chagnon's genealogical data that *unokai* wind up with more extra wives than non-*unokai* and, consequently, more progeny. How the sexual spoils of any raid are divvied up lacks importance to our analysis; we are primarily concerned with the net result of high-status men having greater access. It is unimportant if the object of that access is a homegrown woman or one from outside the tribe.

The Yanomamö Process

In the preceding narrative about the *unokai,* we discussed a simple process for changing the distribution of heritable traits within a population. Among the Yanomamö, achieved status, such as earning the title *unokai,* resulted in increased access to women over those of normal status. This leads to differential reproductive success or fitness, as greater access translates to more mating opportunities. The important consequence of the foregoing is that the Yanomamö population will reflect, over time, the traits of the *unokai.* We will refer to the generic form of this process as the "Yanomamö Process."

Other than status derived from being a successful warrior or the village chief, the Yanomamö are not hierarchal. This lack of social structure is typical of egalitarian, pre-state, tribal societies. Modern, advanced societies are hierarchal and must use a process different from what the Yanomamö use to achieve similar ends. We will refer to this different process as the "English Process." We shall see that for a stratified society, just as for the nonstratified, the question of who gets the girl is important—it just gets more complicated.

The English Process

Data for the English Process come from Clark, who, in *Farewell to Alms,* presents a case of differential reproductive success among the English akin to that of the *unokai* in Chagnon's study of the Yanomamö. In developing his case, Chagnon used tribal genealogy, and for his, Clark uses data from English testators' wills. Clark's data show that rich men had more surviving offspring than poor men did.

For the English, being rich, rather than a killer warrior, granted a reproductive advantage, but for different reasons. Just as the *unokai*'s advantage is predicted to lead to the Yanomamö becoming better at war, the fitness advantage of the English rich is predicted to result in the English becoming better at economic activity.

In a review of Clark's book, Betzig stated that "rich people *always* had more children than poor people, before the Industrial Revolution."[26] According to her argument, the rich leaving more progeny than the poor in preindustrial England, as reported by Clark, is the expected result.

Her argument is correct, but misleading; it lacks completeness. Furthermore, in doing so, it misses the main point of Clark's analysis. The rich having more children than the poor is only a single cog of a more elaborate Darwinian machine. When we parse Clark's words from the following quotation, we will see other cogs, levers, and drive chains. There is more involved than simply being rich and having kids.

The example we studied using the *unokai* as the reproductively advantaged is of a different and simpler process than found among the English. The English case has a larger set of requirements for it to function. The following section from Clark gives us a view of this difference:

> For England we will see compelling evidence of differential survival of types in the years 1250–1800. In particular, economic success

translated powerfully into reproductive success. The richest men had twice as many surviving children at death as the poorest. The poorest individuals in Malthusian England had so few surviving children that their families were dying out. Preindustrial England was thus a world of constant downward mobility. Given the static nature of the Malthusian economy, the superabundant children of the rich had to, on average, move down the social hierarchy in order to find work. Craftsmen's sons became laborers, mechanics' sons petty traders, large landowners' sons small holders. The attributes that would ensure later economic dynamism—patience, hard work, ingenuity, innovativeness, education—were then spreading biologically throughout the population.[27]

This quotation is not simply about rich men and their children; it is much more complicated, as there are many interacting parts. Clark's thesis that the observed process led to traits promoting economic success "spreading biologically throughout the population" gives his book about economics an importance in the biological literature. This is the seed of our analysis and its foundation.

Requirements for the English Process

From the foregoing quote, we can deduce the implied requirements of the English Process:

1. Economic inequality is the primary requirement. Egalitarian societies such as the Yanomamö lack this.

2. A wealth differential must result in higher living standards for the wealthy. Living standards for egalitarian societies are the same.

3. Enhanced living standards must result in the wealthy having more surviving progeny.

4. To effect changes in a population, economic success must be a consequence of heritable traits and not inherited wealth or status. Because no traits are specifically associated with inherited wealth or status, success due to this class has no evolutionary consequence.

5. Heritable traits promoting economic success include diligence, intelligence, and hard work. These traits can be from both genes and the heritable components of culture.

6. Even though economic inequality presupposes a stratified society, stratification is a separate required element.

7. There must be social mobility within the stratified society. Individuals possessing the appropriate traits must be able to become economically successful. This implies that institutions cannot block mobility.

8. Most must be in the game. Changing a population's behavior, be it via genes (allelic frequency) or heritable culture, requires that most be in the game.

9. The key for the English Process is not just mobility but downward mobility. The genes and culture of the scions of the high-status men (the rich in England) have to fill in the slots left by the less genetically and culturally endowed lower-status men or poor for the resulting behavior of the entire population to shift toward that of high-status men.

10. The population size should be static, or nearly so. Although a fixed population may not be necessary for this process to work, it certainly makes the model simpler to understand. Where there is a fixed population and where one group out-reproduces another, mathematics dictates that the scions of the high reproducers fill in for vacancies left behind by the low reproducers.

It is important to note that Clark's data are for pre–Industrial Revolution England. This was a time of high death rates, primarily due to disease. It was also a time of high birthrates. Coupling these with an essentially static population creates the ideal environment for the English Process.

We will discuss later that this process does not readily operate in our modern world because the rich nations both feed and cure disease in the poor ones. Before the Industrial Revolution and with the English Process, the rich out-reproduced the poor because the children of the poor had a higher differential death rate. Almost the opposite is happening today, where the rich have lower birth-rates and the death rates among children of the poor are not as high as they would be without assistance from the rich. That might not have been a nice thing to say, but it is true.

Two Flavors of the English Process

There are two flavors of the English Process. The one we have discussed is simply the rich having more surviving progeny, while the second is a consequence of the first. Just the fact that the rich are fitter makes them attractive to women. Women seek fit men to be the fathers of their children because evolutionary fitness is a predictor of evolutionary success. Women, just as much as men, want their genes to go forward in time. This results in rich men having greater access to women. The first flavor results in more surviving progeny and the second in more progeny.

Rich Chinese men exemplify an example of the second flavor. Just like the English, they have better living standards than poor men do, and they have a "wife and children." Where they differ over their English counterparts is that they also have concubines. These are effectively extra wives, and like extra wives, they produce more progeny for the men who can afford them.

In modern, advanced societies, rich, powerful men have greater access to women than those of lesser status. If you dig down into any "Western" society, I expect that you will find that men of wealth, its perception, or its analog, power, have more progeny or more surviving progeny than those not similarly blessed. Rich men may not have official concubines or live in a polygynous society with extra wives, but they do get more mating opportunities than poor men. Mating opportunities translate into differential reproductive advantage or increased fitness. Rich and powerful men are fitter.

There is a consequence to the rich having more surviving progeny than the poor. This is the Darwinian definition of reproductive advantage and defines the rich as fitter than the poor. The primary consequence of increased fitness is that the heritable traits promoting it will increase in the population. For the case discussed by

Clark, this resulted in the economic prowess of the English population increasing. This is just like the heritable warrior traits of the *unokai* increasing among the Yanomamö. Moreover, just like the Yanomamö becoming the best warriors they can be, the English became the best "economic" society they could be.

The Larger Narrative

We have discussed two processes, the Yanomamö and the English, with each resulting in changed populations. The Yanomamö became better at war and the English better in economic acumen. However, each is only a kernel of a broader, deeper narrative; what we have discussed is simply the part in the very center. Each process has a separate story getting to its middle, and on the opposite side, each has one about its extended consequences. Both are part of the evolutionary machinery at the core of this book. The main point is that they have resulted in the people of this world becoming different and unequal.

CHAPTER 6

Rediscovery of an Important Transition

◇◇◇

In the English Process, the rich are fitter than the poor are, making it a likely candidate for explaining, at least in part, why some nations are rich while others remain poor. This raises the question, what is the source of the English Process? To answer this, we will first have to answer a broader question: how did a people who were probably once similar to the Yanomamö become English? In this chapter, we will start to develop an answer to that question.

We are going to find our answer in a somewhat forgotten transition, one awaiting rediscovery. We will see that its importance is equal to both the Neolithic and Industrial Revolutions. That it has been left lying in the dustbin of seemingly "so yesterday" ideas does not speak well of the academy of social scientists.

Its importance lies in the fact that this is where the path of development of human nature makes a very important change in direction; it is where we seemingly make a U-turn in our evolutionary trajectory. Our societies change from collective, egalitarian, kinship groups to ones with greater individualism. We move away from advanced social societies to ones with greater individualistic nature. We go back toward our old, chimpanzee-like ways.

For the purposes of answering our initial question, this transition separates the rich from the poor; all the rich made the transition. However, we will further see that, as Clark points out (discussed later), not everyone made this transition. For this, there are important consequences, as this is where we find many of the poor of today's world.

A Hidden Transition

In the discipline of anthropology, the transition from tribal, pre-state societies to modern, state-based ones should be a well-understood and established process. It might be, and for anthropologists, it probably is. However, I do not see it, and neither did Gregory Clark. If he had, he would not have written as he did, and this accounting would be significantly different.

In a published response to four separate reviews of his book, Clark speculates about an unknown and unnamed transition in the course of human social development. He develops a case that there is a "hidden transition" between the Neolithic and Industrial Revolutions, one explaining the transition from tribal to state-based societies. The following is the complete quotation of the applicable section:

> For those interested in long run history, the Neolithic Revolution
> looms large. In A Farewell to Alms, this revolution is given very
> much a secondary role to the one great event in history, the Indus
> trial Revolution. My thinking was that settled agriculture was really
> just a natural extension of hunting and gathering. Hunter-gatherers
> before the Neolithic Revolution had been steadily improving their
> efficiency over the course of at least 100,000 years. Indeed arche
> ologists of the Paleolithic have posited their own Upper Paleolithic
> Revolution.[1] Rates of change of technology in the so-called "Neo
> lithic Revolution" were also an order of magnitude lower than in the
> later Industrial Revolution.
>
> The book thus conceives of a gradual evolutionary process.
> Hunter-gatherer societies, with little trade, little capital and much
> violence, give way to settled agrarian societies which are trade and
> capital intensive and more pacific. The evolution of the institutions
> of settled agrarian society—markets, secure property rights, inheri
> tance, limits on violence—is seen as stemming from the technology
> they employed. This through the operations of the Malthusian pro
> cesses leads to a change in human nature and eventually much faster
> economic growth.

Once we follow this logic the tremendous advantage that East Asia and Europe had in technology over Africa, the Americas, and Oceania by 1500 would have to stem from some geographic advantage they had that have [*sic*] led to faster development of settled agriculture in these areas. History would again ultimately be determined by geography, as in Guns, Germs and Steel.[2]

I now feel that this simple picture is incorrect. That there was another important phase transition in history preceding the Industrial Revolution, but NOT the Neolithic Revolution. What gives me pause is the example of settled agrarian societies like the tribes of Papua New Guinea, who operate settled systems of capital intensive agriculture, but still in a social setting where violence is omnipresent, and economic success dependent heavily on family connections, and success in *violence*.

The Huli of Papua New Guinea, for example, who inhabit the southern Highlands, had a settled system of agriculture in 1960 that revolved around sweet-potato gardens, fruit trees, and pig rearing. Wealth is accumulated in the form of pigs mainly, and in control over gardens. Yet this was a society where low-level conflict was endemic. Genealogical interviews suggested that 20% of male deaths (and 6% of female deaths) were from warfare or violence.[3] This is a fantastically high rate compared to societies such as England in the years 1200–1800. The sources of these conflicts were numerous: murders, failure to pay indemnities from earlier wars, breaches of bride price rules, theft, land encroachment, unpaid debts, and rape. Huli behavior in these respects was dominated by impulsiveness. "Huli are volatile and quick tempered; they lack deliberation and they swiftly resort to arms."[4]

Further in Huli society though high social status was associated with wealth, the causation seems to have been that social status produced wealth rather than the reverse. The career of men who became "big men" would typically follow three stages. In adolescence and early manhood, before marriage, they would distinguish themselves in fighting—often with no personal concern with the issues that led to the conflict. In the next stage of their careers, they would begin to accumulate wives, using in part their prestige from their success in fighting. "People esteem successful warriors."[5] To be wealthy meant controlling many gardens and pigs, and this in turn required wives to supply the labor. If they are successful in the various conflicts their affinities and marriages inevitably attach them to, they extend their influence and wealth over a wider area through further marriages. One former war leader who became wealthy and important had married 21 times by age 50.[6]

Similarly Sillitoe reports of Papua New Guinea tribes that though they vary enormously culturally, they are all "big men" societies where "Big men achieve their positions because they excel in the things that matter in life, they are good talkers, they are courageous, they are skillful in exchanges of wealth."[7] Success in these societies—status, pigs, and wives comes not from skill in production or innovation, but from success in war, social intercourse, and social negotiations.

So the move to societies where violence is centralized and limited seems to be an important social evolution distinct from the achievement of settled agriculture. It was this that changed the nature of competition in societies for reproductive success towards strictly economic means. This evolution took place in some parts of the world, but notably not in others such as the Highlands of Papua New Guinea. This suggests that there was another interesting institutional transition much earlier than the Industrial Revolution, and one that we have little understanding of. Economic history, as George Grantham argues, may indeed have three phases as opposed to two.[8]

Clark is asserting that with "an interesting institutional transition," competition for reproductive success changed from one based on "war, social intercourse, and social negotiations" to one based on "economic means." This describes a change from the *unokai* and Yanomamö Process to the English merchant and the English Process. It also describes a transition from *Homo bellicus* to *Homo economicus*. With this unnamed transition, men now compete for access to women with money and power, not with prestige gained as a successful warrior.

Chagnon offers similar insight:

Because intragroup conflicts over local resources (such as females) increase with local group size, continued growth in community size was constantly limited by these conflicts. Such conflicts are most easily resolved by group fissioning making conflicts less likely by reducing the size of the group. However, since there is safety in numbers, there were countervailing pressures to continue to grow. The result was constant pressure for adult males to develop kinship-defined coalitions that peaceably distributed females and pooled their collective abilities to wield force against other groups.

As I see it, the most probable evolutionary scenario in the development of human sociality was constant, but slow growth of community size followed by fissions of communities once they reached sizes of 50 to 75 people—the numbers mentioned in Richard Lee and Irven DeVore's classic treatise, *Man the Hunter.*

It seems that the primary source of conflict within stateless human groups is young females of reproductive age who have neither fathers nor husbands to safeguard them. They constitute a major cause of potential instability and strife among the Yanomamö: aggressive adult males between the approximate age of 15 and 40 constantly want to copulate with them and/or appropriate them for their exclusive reproductive interests. They are a source of conflict in Yanomamö communities and in many other tribally organized societies that lack the institutions associated with the political state—law, police, courts, judges, and odious sanctions.

I suggest that conflicts over means of reproduction—women—dominated the political machinations of men during a vast span of human history and shaped human male psychology. It was only after polygyny became "expensive" that these conflicts shifted to material resources—the "gold and diamonds" my incredulous colleagues alluded to—and the material means of production. By that time, after the agricultural revolution, the accumulation of wealth—and its consequence, power—had become a prerequisite to having multiple mates.[9]

He is suggesting that with the political state comes a basis other than warrior status to compete for access to women. That basis is *"the accumulation of wealth—and its consequence, power."*

With the foregoing, both Clark and Chagnon lend insight to the question, how did a Yanomamö-like tribe become English? They are clearly presenting a case for the shift from *Homo bellicus* to *Homo economicus.* This shift represents a transition from pre-state societies where successful warriors get the woman to state-based ones where wealthy and powerful men become the winners. This transition changes the nature of competition over access to women.

Unmasking the "Hidden Transition"

Clark refers to three institutional transitions in the history of human development. These are the Neolithic Revolution, his *"interesting institutional transition,"* and the Industrial Revolution. His *"interesting institutional transition,"* our model's transition from *Homo bellicus* to *Homo economicus,* and Chagnon's change to the "*political state*" are the same. It is archeologist V. Gordon Childe's Urban Revolution.[10]

The Urban Revolution is not simply about the transition from smaller-scale societies like tribes and chiefdoms to state-level polities or about the formation of urban centers and their characteristics. It is that, of course, but also a great deal more. It is about the social changes that occurred with the Urban Revolution. It is about a major transition in man's evolutionary path. It is about the rise of civilization. For our story, this is where the Yanomamö transition to become English merchants and where the Yanomamö Process shifts to the English Process.

Childe's 1950 précis[11] listed ten characteristics of the Urban Revolution:

1. The urban centers that formed the foundation for the first civilizations were more extensive and had higher population densities than earlier settlements.

2. The composition and function of these urban centers were different from earlier settlements. Although most of the population was concerned with food production, there were specialized full-time non–food producers such as traders and craftsmen. These nonproducers were dependent on the surplus production of the food producers from both the city and the surrounding dependent villages.

3. Food producers paid a tax in the form of excess production to the state, which became available capital. The state could also use this for sustenance for the state's workers.

4. A characteristic of the foundational urban city was large-scale public works represented by monumental public buildings.

5. Society was stratified. There was an elite cadre consisting of rulers, priests, noblemen, military leaders, bureaucrats, scientists, and others who had no role in food production. They provided to the masses certainty of existence (yes, the river will flood again this spring), safety from warring outsiders, rules and their enforcement for ensuring acceptable

social behavior, and freedom from thinking. The masses provided the primary production for the urban enterprise.

6. Management of the urban enterprise required written records for such things as size of the harvest and the amount of tax due. It also required numeracy for accounting. Some level of literacy and numeracy is a characteristic of the urban enterprise.

7. Science, exact and predictive, was also a characteristic. This includes calendars for predicting important events, such as floods, and means of making calculations, such as those required for establishing the batten of a pyramid.

8. Another characteristic was sophisticated art.

9. Long-distance, foreign trade for both necessities, such as ore for making copper, and luxuries, such as finished jewelry for adornment of the elite class, was also a characteristic.

10. With the Urban Revolution came political organization based on residency in the state, not membership in a village. The individual became a citizen and identified with the state. Before, the individual identified with the village or tribe.

Childe's Urban Revolution is not a checklist of things or a discontinuity between two different conditions; it represents a gradual, incremental process. It is the social and cultural changes undergone in changing from kin-based, agricultural villages to complex, urban societies. It is the process of changing from egalitarian, subsistence agriculture to labor specialization in the hierarchal city. There were other changes, but we will learn that for our analysis, the main one was our change in identity from the local kinship group to the large, impersonal state.

Urban Revolution

Introduction. The ascent of urban centers defines the Urban Revolution and is the fountainhead of civilization. Simply stated, the Urban Revolution is the process that resulted in civilization. Urbanization was the enabler, and the evolution of the state polity is an associated consequence. This makes the state a close associate of both cities and civilization. Civilization, the state, the city, are all joined at the hip.

In later chapters, we will spend considerable time on these three major consequences of the Urban Revolution. Before we can do so, though, we must bring into our analysis other threads. So, all we can do now is paint with a broad brush a picture of the nature of the evolution of the Urban Revolution.

Our Yanomamö-like ancestors lived in smallish kinship groups. Each group was self-sufficient, needing little, if anything, from other groups. Within the group, there was little or no specialization, and the basis of division of labor was age and gender. These were egalitarian societies. Once they changed from nomadic foragers to sedentary agriculturists and settled down after the Neolithic Revolution, the people typically selected a chief, or "big man," whose powers often were constrained solely to persuasion. These societies were classical communist organizations; each produced what she could, and all received an equitable distribution. Their ethos was that no one starved unless all starved.

There were two primary issues for survival. One was production of sufficient food, and that was the primary focus of labor. Individuals labored for the good of the group, and the group collective, through its rules of social behavior, ensured that all were fed.

The second primary issue was protection from other human groups. Groups, for varying reasons, attacked other groups, and this occurred on a relatively frequent basis. Sometimes it was for retribution from a previous attack; other times, it could be pure offense for acquiring a needed resource. Kidnapping for women or slaves also entered into the mix. The bottom line was that human groups needed protection from others, and if they did not adequately meet that most basic requirement, they could, and did, become history.

As long as there was parity in power between groups, further political organization for increasing security was not necessary. Change happens, and in time, chiefs became hereditary and multiple groups formed associations such as chiefdoms. This coalescing of groups ended parity and resulted in others having to follow suit or else risk annihilation or absorption by the larger, better-organized groups.

Even though we formed large associations of groups, the individual remained tied to the local group and not to the larger organization. Becoming an individual of a large organization such as a state and not the local group was an identity change waiting in the wings. We first had to invent cities and the state.

The city. It was only after food production capacity increased to the point of excess that the individual could be other than a farmer. On a society reaching that milestone, some could become full-time specialists, such as carpenters, spear makers, and potters. At that time, cities could and did form.

Cities formed and society partitioned. Farmers lived and farmed in the rural periphery while specialists labored in the urban inside. The broader society, just as now, divided into city folk and country folk. In addition, just as now, the city folk required a central government. Self-sufficient and internally reliant, country folk had no such need.

The city brought with it a central authority, the polity of state. This was a hierarchal, central authority whose chiefs and functionaries also lived within the city. There was also a parallel power structure of priests, whose main function was often providing legitimacy to the political elite. We have a city of ruling elites, functionaries to make government work, a military to maintain order, priests to bless, artisans, merchants, and others. The majority remained as peasant farmers. They lived outside the city but produced the wealth to support this new entity.

Before the city, society was egalitarian, but with the city and the state, life became hierarchal. Before, all did all, and the individual labored for the group. Urbanization brought division of labor, and all were incapable of doing all; each could only perform a specialty. Importantly, the individual labored for self. Because different occupations have differing values, this created an economic hierarchy, and such hierarchies are incompatible with the egalitarian group.

The individual now had a place within a hierarchal structure, and the group, at least within the city, lost its reason.

The foregoing may not seem important. However, we will see throughout the story to follow that this is one of the most important changes in our evolutionary path. This is the line in the sand of time where the individual changes identity from the local kinship group to the state. We are going to be spending a lot of time on this topic.

The state. The state comes with the city. It is a political organization with centralized authority and a hierarchal structure. There is a head of state, perhaps a king, and there is often a religious authority providing legitimacy for the ruling class. Others include hereditary elite, a permanent military, and various state functionaries. None of these produces food; they are dependent on the peasants for this.

The first rule of evolution is survival, and the first obligation of the state, after ensuring its own survival, is security for all. Before we had the state, the group met its security requirements by everyone joining in the fray as needed. The state accomplishes these ends with a permanent military for providing security against outside forces. Importantly, from the perspective of survival of the state itself, it uses the same force for self-preservation against internal foes.

The biggest expense of the state is security. To pay for this and its other requisite obligations and duties, taxes are levied. In early civilizations, the only source of tax revenue was production by the peasant farmer. The city and the state could not exist without there being a potential for production in excess of the needs of the producers.

Before the state, the group, acting as a collective, decided everything. It was a true bottom-up system. With the state comes top-down organization, replacing the bottom-up collective. Like the egalitarian collective, the state dictates and enforces the rules for living. The state tells the individual what to do and not to do. Individuals not conforming to the state's wishes receive punishment, just as before.

There is a huge difference between rule by the state and rule by the majority. Owing to self-interest, individuals of the majority, when exercising majority rule, should be fair and equitable. After all, they too could be in the same predicament. The state, controlled by a ruling elite, has no such constraint.

Civilization. Before the Urban Revolution and the advent of city and state, society was simple. It consisted of kinship groups that were essentially closed and self-sufficient. They were self-regulating, as the group was sole arbiter not only of disputes but also of everything social. Such societies had little or no need for government.

Civilization, cities, and the state are almost a package deal—get one, get all three. Social, political, and economic complexity are properties of the city and the state. They are therefore properties of civilization, making it complex.

The following, taken from Wikipedia, is a definition of civilization:

> any complex society characterized by urban development, social stratification, symbolic communication forms (typically, writing systems), and a perceived separation from and domination over the natural environment by a cultural elite. Civilizations are intimately associated with and often further defined by other socio-politico-economic characteristics, including centralization, the domestication of both humans and other organisms, specialization of labor, culturally ingrained ideologies of progress and supremacism, monumental architecture, taxation, societal dependence upon farming as an agricultural practice, and expansionism.

There is a far superior definition in Huntington's *The Clash of Civilizations,* but it is too long for our needs. However, length is not important; context is. Importantly, neither Wikipedia nor Huntington strikes the note we are trying to achieve. Our tune is unique. Ours is about changing identity from the group to the state, and it is on this side of the identity change where we can find civilization. This is a complex topic involving urbanization, states, natural selection, and far more.

For all our history, up until the time we started to form cities, we lived in groups having a high level of genetic relatedness.

There was a genetic shackle coupling us as individuals to a local group of somewhat related individuals. This was a kinship group, and we identified with that group. Eventually, higher-order societies formed, such as chiefdoms and kingdoms, but the basic unit of organization remained the kinship group. This was the group of individual identification.

This relationship changes with the Urban Revolution. In the new urban world, the kinship group loses its value and ceases to exist. The ascendant state can perform all the functions of the old group and, importantly, can meet the requirements of the new, something the old group could never do. On this new, civilized side of history, individuals no longer identify with the local group. In its stead is the broader state. This is where we will find our definition of *civilization*. It is one we will develop throughout the course of our narrative.

Toward the end of our story, we will analyze all eight of our religions/geographies. The four economically worse performing are sub-Saharan Africa, Southeast Asia, Islamic, and Hindu. We will be able to explain the status of the first two as due to remaining tribal and not passing through the portal of the Urban Revolution. As we will have learned, this is the formula for remaining poor.

The last two, the Muslims and Hindus, are outliers. They have some of the oldest civilizations, having undergone the Urban Revolution some millennia ago. However, they did not do it as others; they effectively remained tribal. The Hindu caste system, as we will discuss, essentially froze in place pertinent aspects of tribal society. The Muslims and their Ottoman Empire kept the tribal system by using a unique state structure employing kidnapped Christian boys for state functionaries, a topic we will discuss later. The main point is that, even on becoming civilized, both retained a tribal nature. We will later refer to this as a hybrid civilization. Importantly, those individuals maintained their tribal (or caste) identification, not changing it to the state.

The foregoing provides a start of a definition of civilization. It is a social construct where individuals identify with the state and not some group like a kinship group, tribe, or caste. There is a litmus test. For whom is the individual prepared to die? Is it the local kinship group or the state? This, we will learn, is the dividing line between the rich and the poor. Nations for which the population has entered the portal of the Urban Revolution *and* has lost its tribal nature are civilized.

Social Perspective of the Urban Revolution

Childe was an archeologist, a pedigree reflected in his ten points. Characteristics of the Urban Revolution, such as sophisticated art and monumental buildings, are not on our shopping list. Social changes are, and there are two important ones on Childe's list.

One is the stratification of society. Pre-state societies, while having some relatively small degree of political hierarchy, were primarily socially and economically egalitarian. Their unit of organization was the clan, tribe, village, or a similar entity, and within this unit, individuals were equal. However, differences could and did exist between units.

The English Process requires stratification, and this came with the Urban Revolution. It brought not only further political but also social and economic hierarchy. Economically, there were now princes and paupers along with candlestick makers, merchants, weavers, and everything else in the middle. The political realm had kings, noblemen, tax collectors, army generals, and the king's relatives, along with a whole slew of state functionaries; there were, as always, the majority of subservient masses.

There is a second and more important social change. It is that of individual identity. For millennia, we identified with our local groups, which, not incidentally, contained many of our relatives. Within these kinship groups was where we worked, lived, and raised our families. It was here that our social behavior evolved, resulting in a behavior conforming to the group's wishes. Likewise, the behavior of the group evolved to maximize its evolutionary success with us as a component. We supported the group and the group supported us. We were one with the group.

With the advent of the state, we go from being a member of an egalitarian group to a resident and citizen of the state. We change our identity. Given the immense amount of time we had in our old association, changing identity from our familiar kinship group to the large impersonal one of the state was major.

Of all the consequential changes of the Urban Revolution, this is the most important. Those who identify with the state are civilized, and those whose identity remains with the tribe are not. This is an unusual definition for civilization, but as we develop this theme in the following chapters, its truth should become obvious.

At its core, the Urban Revolution is the transition from society organized around small, egalitarian kinship groups to one organized around individual citizens of the large, impersonal state. From this transition cascades a large series of associated changes, especially to human nature. In much of the remainder of our story, we will discuss them.

Cities Change Evolution's Environment

The Urban Revolution gave us civilization, cities, and the state, topics about which anthropologists write reams. They are deserving topics, and we will spend some time on them. There is another discipline, sociobiology, having more than a casual interest in the nature of the creations of the Urban Revolution. As we will discover, this revolution markedly changes the environment for human evolution and becomes a demarcation point on the path of human development.

We could define the city by size, population density, and the other usual traits, but such a classification is not useful for our analysis. The main point is that the city is not a pre-state social group, such as a tribe, village, or assemblage of kinship groups. It is a population of individuals, and once that condition is met, we can address other attributes. The city does not *a priori* preclude groups, but its organizing unit is the individual; groups, if any, are secondary.

The importance of cities to our analysis is not so much in what they are; rather, it is in their consequences. Cities created a new, important arena for evolutionary competition. Before, evolutionary competition between individuals was confined to within the local group. These local groups were not only egalitarian; they proactively enforced egalitarian behavior. Such action discourages competition between individuals, and in this type of environment, evolutionary competition is highly constrained. If everyone exhibits the same behavior, few differences exist on which natural selection can act.

With urbanization, the arena for evolutionary competition changed from within the smallish, kinship group to within the much larger, impersonal city. Leaving the warmth and security of the group forced individuals to fend for themselves and their families. The city offered more opportunity but no succor. Primarily, the city replaced the collective of the old kinship group with the individual, an entity that did not truly exist before.

The city is a complex environment where the individual can rise or fall economically. Within the city, economic competition between individuals becomes a new arrow in Darwin's quiver, as individuals can rise or fall on their own merits. Before, in the egalitarian group, rising or falling would be an oxymoron; there would be no such thing allowed, as it would lead to inequality. In comparison with the group, the city offers both increased risk and potential reward.

Consequence of Not Making the Transition

We may not know all the details or exactly why, but we do know that the Urban Revolution happened. Cities formed, and so did states and civilizations. We do know that it happened at different times for different people, with the early adopters starting before 3000 BCE. Others, like the Western Europeans, did not make the final transition until the Middle Ages some four thousand years later. This transition has only recently started in sub-Saharan Africa and Southeast Asia; there, it is still in progress. Then, there are those, such as the Arabs, who remain tribal and have not made a complete transition. Additionally, for some relatively few, such as the Yanomamö, the transition has yet to start.

The Urban Revolution represents the transition to civilization from primitive societies. Toward the end of our story, when we evaluate our eight religions/geographies, we will have in hand evidence demonstrating that becoming civilized is the principal dividing line between the rich haves and the poor have-nots. The antecedents of the haves are those societies having civilized roots, those making the transition of the Urban Revolution. Populations where the transition from *Homo bellicus* to *Homo economicus* did not occur, or only did so recently, is where we find the poor. It is important to note that having civilized roots is a necessary but not sufficient condition for achieving economic prosperity. India and Egypt are examples.

CHAPTER 7

Group Selection
and the Superorganism

In the prior chapter, we stated that becoming civilized was the first step on the path to becoming rich. It starts with the transition of the Urban Revolution, and those not taking this step remained uncivilized and poor. If our interest were only to understand why the rich came to be economically prosperous, we could start at this point of demarcation. We would have no need to examine who we were before.

As it turns out, our interest is more with the poor. After all, the rich can take care of themselves, and the poor are with whom problems lie. We therefore need to understand who we were before the age of urbanization; once having this understanding, we will be able to reason why the poor are the way they are.

Before the Urban Revolution and for almost all of our time as humans, the accepted teaching of anthropologists was that we lived in kinship groups. This is a correct but not complete descriptor. It lacks an important modifier, one not grounded in anthropology but rather in sociobiology. It is the kinship group as a superorganism.

In biology, an example of a superorganism is a beehive. A human superorganism is similar, being a group, typically of related individuals, that behaves as if it were an organism. Just as cells do for an individual organism, its several individuals do for their superorganism. Primarily, they labor in behalf of their group and not for self. In return, the superorganism meets their needs, with the group's social rules controlling this internal functioning. Although individuals can leave and new ones enter, it is primarily a closed system.

The superorganism, as a sociobiological entity, has several unique and important properties, many of which, such as altruistic behavior, anthropologists never addressed. Additionally, many play directly into the narrative of becoming rich or staying poor. For this, we need to understand them and, to do so, move our analysis from the world of anthropology to that of sociobiology.

Before getting to the superorganism, we first need to look at its underlying cause. In Darwinian natural selection, there is more than one possible target of selection. It can be the gene, the individual, or the group. In the case where the human group is the target of selection, that group is a superorganism. Consequently, we need to start at group selection.

Serendipity under the Rock

Before getting to group selection and thereby change the perspective of our analysis from anthropology to sociobiology, it will pay to relate how I arrived at this point. In developing this part of the analysis, the jump from anthropological kinship group to sociobiological superorganism and then to group selection was not a simple, obvious transition. It was analogous to the serendipitous act of finding a lost class ring and not the anticipated pack of scurrying crabs on turning over a rock in a tide pool at the ocean's edge. It was both unexpected and exciting. The story of this abrupt transition follows.

Understanding pre-state societies is the standard fare of anthropologists and should not be a difficult task. This might seem to be a strange statement, but fortunately, it was difficult. Being so led to an unintended, novel, and very illuminating discovery.

While studying the economics of pre-state societies, an economic text having an appendix containing a dustbin of ethnographic observations was uncovered. The fortuitous nature of this discovery is that these observations led to the conclusion that pre-state societies, or at least some of them, operated under a form of Darwinian natural selection termed *group selection* and were thus superorganisms.

Until somewhat recently, group selection has been a discarded and discredited model of natural selection. In the first part of the last century, the accepted view was that natural selection was "for the good of the species" and group selection ruled. Later, individual selection became the accepted paradigm, as gene-centric models offered solid explanations of observations. With this shift, ridicule and worse became the reward for any out-of-step advocates of group selection. Science does not stand still, and currently group selection, as part of a theory of multilevel selection, is again becoming an acceptable argument in the debate.

Given the disfavor of group selection, it is not surprising that no one considered group selection for human evolution; it has not been addressed. However, given the compelling evidence in the referenced ethnographic observations, this is exactly what we are going to do.

If there is a new scientific point to our narrative, it is that of group selection and the evolution of human nature. Taking this route is not simply the path less traveled; it is the path never ventured. We will find this novel path to be rewarding on more than a single facet. For one, it offers a clear and simple explanation for altruism. On another, it provides a model for understanding the evolution of morals. There is more.

Being the first down a path can have its rewards, especially with discovery of low-hanging fruit. However, such paths can have unrecognized risks, as novel, untested explanations are usually less than robust. They have a greater probability of being incorrect and can even be misleading. Therefore, even though *carpe diem* seemed appropriate when finding these observations, *caveat emptor* is by far the more appropriate response.

The focus of the first part of this chapter is on the observational evidence suggesting that for a large part of our evolutionary history, group, and not individual, selection was the *modus operandi* for human evolution, defining our pre-state groups as superorganisms. The key to understanding who we were and what we have become is tightly bound to this concept.

If group selection was in fact a factor in human development, there must have been a reason. There is. It is the interaction of war with egalitarian societies, the topic of the second half of this chapter.

Introduction to a Debate

Science is not always dispassionate or objective. Science is done by humans, humans with egos, agendas, and all sorts of other psychological baggage. The topic of this part of our analysis, group selection, has been contentious, and the debate illuminates these problems. In some parts of academia, even discourse on group selection is taboo. Brewer and Caporeal bemoan this sentiment: "a long standing prohibition on group selection, which appears to ban any discussion of the role of groups in human evolution whatsoever."[1]

Even though group selection has not been in the mainstream, it does have several respected proponents. The impressive list includes Christopher Boehm, E. O. Wilson, D. S. Wilson, Elliott Sober, Peter Richerson, Robert Boyd, Martin Nowak, Jonathan Haidt, M. B. Brewer, and L. R. Caporeal, all highly respected researchers in their respective scientific disciplines.[2]

The debate about group selection is concerned with a larger question. In natural selection, what is the unit or target of selection? Is it the individual, the group, the gene, or all of the above (multilevel)? The following is about that debate.

Before the 1960s, the understanding was that evolution functioned at the group level. Natural selection operated on the group with evolution being for the "good of the species." In the 1960s, that model changed with the development of a gene-centric theory, with R. A. Fisher in 1930 and J. B. S. Haldane in 1932 and 1955 laying the theories foundation.[3] It was further developed and popularized in 1964 by W. D. Hamilton,[4] and then, in 1976, Richard Dawkins's popular book *The Selfish Gene* put an end to any consideration of group selection. The gene became the accepted unit of selection with the individual as its target.

A major problem with the old group selection theory was that it did not explain altruism. This is when an individual acts in a manner that decreases the individual's fitness but increases the fitness of the recipient of the altruistic act, whether it be another individual or a group. Cutting the grass of a sick neighbor is a trivial example; dying in war for your tribe or country is not.

The new model, *inclusive fitness,* is based on the fact that related individuals share some identical genes. From the genes' point of view, it makes perfect sense for an individual to act in a manner that decreases the individual's fitness while increasing the fitness of close relatives; the genes win no matter what. The genes' only evolutionary interest is leaving behind a maximum number of copies of themselves in the population.

With inclusive fitness, genes, not individuals or groups, are the unit of natural selection. Genes survive, reproduce, and have a permanence much longer than their carriers, individuals, have. It is the relative temporal permanence of the gene over the individual or group that makes it the effective unit for selection. The code of the gene spans innumerable generations; the term of the individual is only one. This gene-centric perspective became widely accepted and group selection fell by the wayside, discarded and discredited.

Dawkins's book, *The Selfish Gene,* introduced some powerful metaphors into the vocabulary of evolutionary biology, one being the *selfish gene.* Genes, of course, cannot be selfish, but we can metaphorically conceptualize them as such. Another is *survival machine.* This concept is that we, like individuals of any species, are the survival machines for our genes. Dawkins succinctly states that "we are survival machines—robot vehicles blindly programmed to preserve the selfish molecules known as genes."[5]

The expression itself is not important to our conversation; it is the color of the metaphor. We are not concerned with the term being correct or not. Its visceral nature will sharpen our understanding and shepherd our discussion.

Science advances and group selection is not dead. Sober and Wilson's 1998 book *Unto Others* brought it back to life, where it becomes a part of a broader theory: multilevel selection. Multilevel selection theory posits that natural selection can operate at more than just a single level of the biological hierarchy from the molecule (gene) to group. The new theory includes the group, individual, and gene as possible targets of selection, with the gene remaining as the unit of selection. Selection chooses the most adaptive traits, whether the expressed trait is at the group, individual, or cellular level of the biological hierarchy. Genes are the messengers between generations.

Edward O. Wilson is one of the preeminent biologists of our time, and in his book *The Social Conquest of Earth*,[6] he came down on the side of group selection. This added fuel to the debate's fire, stirring up a big pot in academia. What is important is that he lent respect to group selection.[7]

With the new multilevel selection scheme, some, but not all, in the scientific community accepted group selection within that context. However, the debate continues. Some of it is due to "a massive confusion between alternative theories that invoke different processes, on the one hand, and alternative perspectives that view the same process in different ways, on the other."[8]

There are some large egos involved. With process confused with perspective, conversations will continue talking past one another for a time.

Asking the Critical Question

In the following, we are going to examine some ethnographic observations, and informing our understanding of the observations will be how we answer the question, what is the target or unit of natural selection?

The observations we will examine are mainly about work attitudes in tribal societies. The following provides a reference condition we can use in arriving at a conclusion:

> We should expect an individual as a survival machine to exhibit behavior promoting its survival and reproduction, not that of the group.

Are the observations consistent with the group or the individual as the *survival machine*?

The following observations come from Bronisław Malinowski's well-known 1922 ethnographic study with the wonderful title *Argonauts of the Western Pacific.*[9] The Argonauts are the people of the Trobriand Islands, an island chain located to the northeast of New Guinea:

> He works prompted by motives of a highly complex, social and traditional nature, and towards aims which are certainly not directed towards the satisfaction of present wants, or to the direct achievement of utilitarian purposes. Thus, in the first place, as we have seen, work is not carried out on the principle of the least effort. On the contrary, much time and energy is [*sic*] spent on wholly unnecessary effort, that is, from a utilitarian point of view. Again, work and effort, instead of being merely a means to an end, are, in a way an end in themselves. A good garden worker in the Trobriands derives a direct prestige from the amount of labour he can do, and the size of garden he can till. The title *tokwaybagula,* which means, "good" or "efficient gardener," is bestowed with discrimination, and borne with pride. Several of my friends, renowned as *tokwaybagula,* would boast to me how long they worked, how much ground they tilled, and would compare their efforts with those of less efficient men. When the labour, some of which is done communally, is being actually carried out, a good deal of competition goes on. Men vie with one another in their speed, in their thoroughness, and in the weights they can lift, when bringing big poles to the garden, or in carrying away the harvested yams.
>
> The most important point about this is, however, that all, or almost all the fruits of his work, and certainly any surplus which he can achieve by extra effort, goes not to the man himself, but to his relatives-in-law. Without entering into details of the system of the apportionment of the harvest, of which the sociology is rather complex and would require a preliminary account of the Trobriand kinship system and kinship ideas, it may be said that about three quarters of a man's crops go partly as tribute to the chief, partly as his due to his sister's (or mother's) husband and family.
>
> But although he thus derives practically no personal benefit in the utilitarian sense from his harvest, the gardener receives much praise and renown from its size and quality, and that in a direct and circumstantial manner. For all the crops, after being harvested, are displayed for some time afterwards in the gardens, piled up in neat, conical heaps under small shelters made of yam vine. Each man's harvest is thus exhibited for criticism in his own plot, and parties of

natives walk about from garden to garden, admiring, comparing and praising the best results. The importance of the food display can be gauged by the fact that, in olden days, when the chief's power was much more considerable than now, it was dangerous for a man who was not either of high rank himself, or working for such a one, to show crops which might compare too favourably with those of the chief. . . .

He is not guided primarily by the desire to satisfy his wants, but by a very complex set of traditional forces, duties and obligations, beliefs in magic, social ambitions and vanities. He wants, if he is a man, to achieve social distinction as a good gardener and a good worker in general.[10]

Answering the Question

The Trobriand Islander does not receive any reward other than praise for his work. He works for prestige and social status, not to fill his belly. This is not the behavior of a survival machine; survival machines work to fill their and their families' bellies. A Dawkins survival machine works directly for its genes' survival and reproduction, not that of the group.

Malinowski's observation makes sense, but *only* if the survival machine is the group, not the individual or the individual's genes. We can model the following for Malinowski's Trobriand Islanders:

> It is the group, not individuals or their genes, that are the target of natural selection. The survival machine is not the individuals of the group; it is the group, with the genes of the individuals acting as its messenger.

The individual as survival machine is simple—just look in the mirror. In modern man, the individual is the survival machine. What does the survival machine look like when it is the group? What does the group see when it looks in the mirror?

The human group, as survival machine, is not an idea found in texts; we will have to synthesize our own. Do not be concerned that we are going off the intellectual deep end. We are going to find plenty of indirect support for this concept, especially when we examine pre–Urban Revolution societies from that perspective.

The individual as survival machine is not simply just an individual; it is a collection of cells organized to function as a cohesive unit of survival and selection. The group as survival machine is similar, just one level higher in the biological hierarchy. The group survival machine consists of individuals organized to function as a cohesive unit of survival and selection.

This group is like an organism. In fact, we are going to call the group of group selection a "superorganism." Our human superorganism is the target of group natural selection.

The term *superorganism* is often used to describe a colony or hive of social insects such as ants, termites, and bees. We get a definition from Kesebir:

> Biologists call highly cooperative and socially integrated animal groups such as beehives and ant colonies—superorganisms. In such species, the colony acts like an organism despite each animal's physical individuality.[11]

The same concept applies to the human superorganism. It is a social group of individuals acting collectively as if they were an organism—a superorganism.

Superorganisms function to meet the needs of the collective of its individuals, and its individuals, in turn, behave to fulfill that mandate. For this, social insects use genetically determined rules, whereas humans use, at least in part, culturally determined ones.

A metric for determining if a human group is a superorganism is if its individuals labor primarily for good of the group and not self. This requirement defines the superorganism as being populated with altruists, and the preceding observations along with those that follow suggest that this is true. Altruism and the superorganism is an important topic, one that we will later discuss.

It is as a cohesive group that the superorganism survives, or not, determining the fate of its individuals. The interests of the superorganism and its individuals are aligned, making this arrangement the classic case of "all for one and one for all."

Superorganisms are not short-term, ad hoc groups like a fan club or a baseball team. They have stability over extended time, spanning generations.

The human superorganism is egalitarian, but it does not have to be. However, as part of the process of acquiring group selection, it became egalitarian, but not purely so. There are two classes, male and female, and as some have yet failed to recognize, they are not equal. Additionally, we will see that some individuals achieve high status for superior effort on behalf of the superorganism and thereby become unequal.

In theory, the human superorganism does not have a chief, but in practice, ever since the Neolithic Revolution, there has usually been one. Becoming chief is usually based on merit and is not hereditary. Often the basis for selection as chief is having excellent persuasive powers. In any event, the chief had little power and was usually *primus inter pares*. We will later discuss that we started to lose our superorganismic ways when chiefs became hereditary.

Our group selection machine or superorganism is not a strange, unfamiliar creature. We have already examined one, the Yanomamö. Others are the several thousands, perhaps millions, of similar societies that existed in pre–Urban Revolution times.

This is a bare-bones model of group selection and the superorganism. Over the following narrative, we will be adding a lot of flesh, but for now, we have enough to go forward.

Supporting Observations

Our speculative model for human group selection currently rests on a single ethnographic study. Be assured, we are not going to leave it there on such a flimsy base. Piecemeal and one by one, we could refer to other similar studies and strengthen our case. This pedantic exercise would show that our single example is only one of many similar observations made by various anthropologists.

Fortunately, there are several applicable ethnographic observations located in a single book. Interestingly, it is an economic, not anthropological, text, but in any event, it relieves us of the systematic, arduous task of one-by-one itemization.

The book is Karl Polanyi's *The Great Transformation: The Political and Economic Origins of Our Time*, a well-known work of economic history and social theory. Polanyi used ethnographic data of pre-civilized man in developing his thesis on economics and these same data pertain directly to our narrative.

Normally quotations from a referenced work are used to provide support for a point to be made. Our use of *The Great Transformation* will be slightly different; it will use Polanyi's "Notes on Sources" located at the end of his book. The support for our point comes from Polanyi tying together quotations from several noted anthropologists, such as Firth, Malinowski, and Thurnwald, and they are the source of our support. However, by organizing, Polanyi did all the work.

Read the following long quotation within the context of our model of the superorganism and group selection. Specifically, answer the question of the individual versus the superorganismic group as the target of natural selection: which one is the survival machine?

Selected References to "Societies and Economic Systems"

The nineteenth century attempted to establish a self-regulating economic system on the motive of individual gain. We maintain that such a venture was in the very nature of things impossible. Here we are merely concerned with the distorted view of life and society implied in such an approach. Nineteenth century thinkers assumed, for instance, that to behave like a trader in the market was "natural," any other mode of behavior being artificial economic behaviour—the result of interference with human instincts: that markets would spontaneously arise, if only men were let alone; that whatever the desirability of such a society on moral grounds, its practicality, at least, was founded on immutable characteristics of the race, and so on. Almost exactly the opposite of these assertions is implied in the testimony of modern research in various fields of social science such as social anthropology, primitive economics, the history of early civilization, and general economic history. Indeed, there is hardly an anthropological or sociological assumption—whether explicit or implicit—contained in the philosophy of economic liberalism that has not been refuted. Some citations follow.

(a) The motive of gain is not "natural" to man.

"The characteristic feature of primitive economics is the absence of any desire to make profits from production or exchange." (Thurnwald, *Economics in Primitive Communities*, 1932, p. xiii). Another notion which must be exploded, once and forever, is that of Primitive Economic Man of some economic textbooks" (Malinowski, *Argonauts of the Western Pacific*, 1930, p. 60). "We must reject the Ide-

altypen of Manchester liberalism, which are not only theoretically, but also historically misleading." (Brinkman, Das soziale System des Kapitalismus." In *Grundriss der Sozialdkonomik*, Abt. IV, p. 11).

(b) To expect payment for labor is not "natural" to man.

"Gain, such as is often the stimulus for work in more civilized communities, never acts as an impulse to work under the original native conditions." (Malinowski, *op. cit.,* p. 156). "Nowhere in uninfluenced primitive society do we find labor associated with the idea of payment." (Lowie, "Social Organization," *Encyclopedia of the Social Sciences,* Vol. XIV, p. 14). "*Nowhere* is labor being leased or sold." (Thurnwald, *Die menschliche Gesellschaft,* Bk. III, 1932, p. 169). "The treatment of labor as an obligation, not requiring indemnification . . . is general." (Firth, *Primitive Economics of the New Zealand Maori,* 1929). "Even in the Middle Ages payment for work for strangers is something unheard of." "The stranger has no *personal* tie of duty, and, therefore, he should work for honor and recognition." Minstrels, while being strangers, "accepted payment, and were consequently despised." (Lowie, *op. cit.*).

(c) To restrict labor to the unavoidable minimum is not "natural" to man.

"We cannot fail to observe that work is never limited to the unavoidable minimum but exceeds the absolutely necessary amount, owing to a natural or acquired functional urge to activity." (Thurnwald, *Economics,* p. 209). "Labor always tends beyond that which is strictly necessary." (Thurnwald, *Die menschliche Gesellschaft,* p. 163).

(d) The usual incentives to labor are not gain but reciprocity, competition, joy of work, and social approbation.

Reciprocity: "Most, if not all economic acts are found to belong to some chain of reciprocal gifts and counter gifts, which in the long run balance, benefiting both sides equally. . . . The man who would persistently disobey the rulings of law in his economic dealings would soon find himself outside the social and economic order— and he is perfectly well aware of it." (Malinowski, *Crime and Custom in Savage Society,* 1926, 99. 40–41).

Competition: "Competition is keen, performance, though uniform in aim, is varied in excellence. . . . A scramble for excellence in reproducing patterns." (Goldenweiser, "Loose Ends of Theory on Individual, Pattern, and Involution in Primitive Society." In *Essays in Anthropology,* 1936, p. 99). "Men vie with one another in their speed, in their thoroughness, and in the weights they can lift, when bringing big poles to the garden, or in carrying away the harvested yams" (Malinowski, *Argonauts,* p. 61).

Joy of work: "Work for its own sake is a constant characteristic of Moari industry." (Firth, "Some Features of Primitive Industry," *E.J.,* Vol. I, p. 17). "Much time and labor is given up to aesthetic purposes, to making the gardens tidy, clean, cleared of all debris; to building fine, solid fences, to providing specially strong and big yam-poles. All these things are, to some extent, required for the growth of the plant; but there can be no doubt that the natives push their conscientiousness far beyond the limit of the purely necessary." (Malinowski, *op. cit.,* p. 303).

Social approbation: "Perfection in gardening is the general index to the social value of a person." (Malinowski, *Coral Gardens and Their Magic,* Vol. II, 1935, p. 124). "Every person in the community is expected to show a normal measure of application." (Firth, *Primitive Polynesian Economy,* 1939, p. 161). "The Andaman Islanders regard laziness as an antisocial behavior" (Radcliffe-Brown, *The Andaman Islanders*). "To put one's labor at the command of another is a social service, not merely an economic service." (Firth, *op. cit.,* p. 303).

(e) Man the same down the ages.

Linton in his *Study of Man* advises caution against the psychological theories of personality determination, and asserts that "general observations lead to the conclusion that the total range of these types is much the same in all societies, ft. . [*sic*] In other words, as soon as he (the observer) penetrates the screen of cultural difference, he finds that these people are fundamentally like ourselves" (p. 484). Thurnwald stresses the similarity of men at all stages of their development: "Primitive economics as studied in the preceding pages is not distinguished from any other form of economics, as far as human relations are concerned and rests on the same general principles of social life." (*Economics,* p. 288). "Some collective emotions of an elemental nature are essentially the same with all human beings and account for the recurrence of similar configurations in their social existence." (Sozialpsychische Ablaufe im Volkerleben, in *Essays in Anthropology,* p. 383). Ruth Benedict's *Pattern of Culture* ultimately is based on a similar assumption: "I have spoken as human temperament were fairly constant in the world, as if every society a roughly similar distribution were potentially available, and, as if the culture selected from these, according to its traditional patterns, had moulded the vast majority of individuals into conformity. Trance experience, for example, according to this interpretation, is a potentiality of a certain number of individuals in any population. When it is honored and rewarded, a considerable proportion will achieve or simulate it . . ." (p. 233). Malinowski consistently maintained the same position in his works.

(f) Economic systems, as a rule, are embedded in social relations; distribution of material goods is ensured by noneconomic motives.

Primitive economy is "a social affair, dealing with a number of persons as parts of an interlocking whole." (Thurnwald, *Economics,* p. xii). This is equally true of wealth, work, and barter. "Primitive wealth is not of an economic but of a social nature." (*ibid.*). Labor is capable of "effective work," because it is "*integrated into an organized effort by social forces.*" (Malinowski, *Argonauts,* p. 157). "Barter of goods and services is carried on mostly within a standing partnership, or associated with definite social ties or coupled with a mutuality in non-economic matters." (Malinowski, *Crime and Custom,* p. 39).

The two main principles which govern economic behavior appear to be reciprocity and *storage-cum-redistribution:*
"*The whole tribal life is permeated by a constant give and take.*" (Malinowski, *Argonauts,* p. 167). "Today's giving will be recompensed by tomorrow's taking. This is the outcome of the principle of reciprocity which pervades every relation of primitive life. . . ." (Thurnwald, *Economics,* p. 106). In order to make such reciprocity possible, a certain "duality" of institutions or "symmetry of structure will be found in every savage society, as the indispensable basis of reciprocal obligations." (Malinowski, *Crime and Custom,* p. 25). "The symmetrical partition of their chambers of spirits is based with the Banaro on the structure of their society, which is similarly symmetrical." (Thurnwald, *Die Gemeinde der Bdnaro,* 1921, p. 378).

Thurnwald discovered that apart from, and sometimes combined with, such reciprocating behavior; the practice of storage and redistribution was of the most general application from the primitive hunting tribe to the largest of empires. Goods were centrally collected and then distributed to the members of the community, in a great variety of ways. Among Micronesian and Polynesian peoples, for instance, "the kings as the representatives of the first clan, receive the revenue, redistributing it later in the form of largesse among the population." (Thurnwald, *Economics,* p. xii). This distributive function is a prime source of the political power of central agencies. (*ibid.,* p. 107).

(g) Individual food collection for the use of his own person and family does not form part of early man's life.

The classics assumed the pre-economic man had to take care of himself and his family. This assumption was revived by Carl Buecher in his pioneering work around the turn of the century and gained wide currency. Recent research has unanimously corrected Buecher on this point. (Firth, *Primitive Economics*, pp. 170, 268 and *Die menschliche Gesellschaft*, Vol. III, p. 146; Herskovits, *The Economic Life of Primitive People*, 1940, p. 34; Malinowski, *Argonauts*, p. 167, footnote).

(h) Reciprocity and redistribution are principles of economic behavior that apply not only to small, primitive communities but also to large and wealthy empires.

"Distribution has its own particular history, starting from the most primitive life of the hunting tribes." ". . . The case is different with societies with a more recent and more pronounced stratification. . . ." "The most impressive example is furnished by the contact of herdsmen with agricultural people." ". . . The conditions in these societies differ considerably. But, the distributive function increases with the growing political power of a few families and the rise of despots. The chief receives the gifts of the peasant, which have now become 'taxes,' and distributes them among his officials, especially those attached to his court."

"This development involved more complicated systems of distribution . . . All archaic states—ancient China, the Empire of the Incas, the Indian kingdoms, Egypt, Babylonia—made use of a metal currency for taxes and salaries but relied mainly on payments in kind stored in granaries and warehouses . . . and distributed to officials, warriors, and leisured classes, that is, to the non-producing part of the population. In this case distribution fulfills an essential economic function." (Thurnwald, *Economics*, pp. 106–8).

"When we speak of feudalism, we are usually speaking of the Middle Ages in Europe. . . . However, it is an institution, which very soon makes its appearance in stratified communities. The fact that most transaction are in kind and that the upper stratum claims all the land or cattle, are the economic causes of feudalism . . ." (*ibid.*, p. 195).[12]

In the preceding observations, we see that labor was for the good of the group, not for the good of the laborer. This is the primary determinate of being a superorganism or not and of group selection. In both, the superorganism and group selection, the individual's labor must be primarily for the good of the group, not the individual. This stipulation requires that the totality of that labor along with the social rules for the distribution of labor's fruit meet the needs of all the individuals within. That is the only way this system works.

The preceding quotations highlight the fact that individuals worked for honor, recognition, and social approbation. The motive for labor was not the job; it was to gain status.

This reward of status happens to solve an evolutionary problem, and in an interesting way. Laboring for other than yourself and your own genes decreases fitness, making such altruistic labor maladaptive. Rewarding status to laborers, especially like the *tokwaybagula* of the Trobriand Islands, solves this dilemma. Often, the consequence of earning status is increased fitness, as high-status individuals usually have more and better mate choices—work for the group and lose fitness but gain it back and then some with group-rewarded status.

Evolution is competitive. In the usual individual selection process, individuals compete against one another. The same happens when the individuals are a part of a superorganism, the unit of group selection. In this case, labor is not simply about status; it is about competing for status and winning reproductive advantage or fitness. Individuals compete using labor as a weapon, with the best laborer achieving the highest status. The title *tokwaybagula* is all about competing to be the best gardener and harvesting status as reward.

If the sole purpose of labor were to accomplish a task, once completed, labor would end. However, in the preceding observations, labor is always more than that required. This is a consequence of status being the motive for labor and not the job itself. Moreover, there could be cases where labor has no meaningful purpose other than status.

As noted in the observations, storage and redistribution of food is a feature of pre-state societies. This is an expected behavior for a group survival machine, as a society that stores excess food against a rainy day should have better survival prospects than those that do not.

Reciprocity is another social aspect of labor. In pre-state societies, there was a socially established *quid pro quo* when one labored for another. A person freely worked for others with the social understanding that, when required, the effort would be reciprocated.

We are going to spend more time developing the concept of the human superorganism. However, this brief discussion of ethnographic observations is a sufficient beginning and starts to show that these societies were not only kinship groups; they were more complex.

The Unfamiliar Superorganism

In the world in which I have lived for some time, individuals get up in the morning and go to work. They do so for the selfish purpose of survival of self and family. In an earlier land of the superorganism, the individual labored not for self but rather unselfishly for the group. In return for performing this socially mandated labor, the group functioned to satisfy the needs of its constituent individuals.

I have a perceptual problem with this proposition; it is outside my normal frame of reference. It forces a shift in my frame of reference for work-related behavior. I find it difficult to conceptualize living in a society where I am dependent on a collective for survival and where my only obligation is to do my socially assigned task within the colony. Within this communal system, I have no control over my destiny; I lack any sense of agency. I am only one of a larger collective functioning as an organism, one that has its own destiny. If my group does something stupid, natural selection deselects it, and I become history. All the needs for my family and me come from my superorganism. This is not who I am. This is a severe case of forest–tree syndrome. As I stand as a tree in this forest, it is difficult to understand the forest. However, we will try.

An encompassing metaphorical fence contains our pre-state survival machine, the unit of natural selection. Inside the group is the world as described by Adam Smith:

> And hence it is, that to feel much for others and little for ourselves, that to restrain our selfish, and to indulge our benevolent affections, constitutes the perfection of human nature; and can alone produce among mankind that harmony of sentiments and passions in which consists their whole grace and propriety.[13]

However, the world outside of the group is Hobbesian:

> and which is worst of all, continual fear, and danger of violent death; and the life of man, solitary, poor, nasty, brutish, and short.[14]

On the inside of the fence, we find the gardeners of the Trobriand Islands and the warriors of the Yanomamö. This is not where we find the English merchant. He identifies with the state and not with some tribe.

Inside the fence, man is only a single unit of a group. An analogy is that of the cells of an organism where all function together for the greater good of the collective, the organism. It is in this sense that the individual is part of what we refer to as a superorganism. Importantly, this superorganismic group of individuals enclosed by our fence is the unit of natural selection.[15]

The superorganism defines our nature before the Urban Revolution and, therefore, the nature of those not making its transition. We have stated that the poor of today's world are those not making the transition, and therefore the nature of the superorganism defines the nature of the poor. Additionally, because we were all at one point in this condition, the superorganism defines the starting point for those who are now rich. For good reason, starting with the next chapter, we are going to spend a lot of time understanding the nature of this sociobiological entity.

Egalitarianism, Fountainhead of the Superorganism

In the next chapter, we are going to explore the nature of the human superorganism. Before we do, we should try to understand its roots, and that is what we will be doing in the rest of this chapter.

If we took on group selection and our groups became superorganisms, there must have been a reason. Our evolutionary history must contain the explanation. It does, and we will find the source for many of our answers in Christopher Boehm's book *Hierarchy in the Forest: The Evolution of Egalitarian Behavior.* Boehm develops a model of our transition from chimpanzee-like primates living in some ancient African forest to egalitarian foragers. On examining his model, we will find that the evolutionary mechanics of our becoming egalitarian provide the explanation for why our societies acquired the property of group selection and concurrently became superorganisms.

Our modern Western world is hierarchal, not egalitarian. Attesting to this are the inequalities expressed in the social, economic, and political properties of our institutions. However, when we go back to the time of our forager and tribal ancestors, we find these same institutions to be exemplars of equality. We, along with our institutions, have changed.

To understand this relatively recent change, we need to go back to a time very early in our history, to a place before we were egalitarian. To get where we need to go, we must journey all the way back to a period where we were much more similar to our genetic cousins, the chimpanzees, than we are today.

How far back in time must we trek? Boehm has it before our migration out of Africa, and that was approximately fifty thousand years ago.[16] The exact timing is not important; it does not affect our arguments. It probably lies between a hundred thousand and a million years or so. Keeley has human existence as 2 million years.[17] If we need a date certain, we will assume 1 million years.

At this starting point, we were foragers in Africa, living in relatively small hierarchal bands or groups of a few tens of individuals. Back then, we were just like the other primates, hierarchal in the forest with despotic alpha males at the top.

Males competed for status, with the winning competitor becoming the dominant alpha male. He got the reproductive rights to women and, by that measure, had greater fitness than other males.

Boehm posits a conflicting duality to our nature. It was more than men simply wanting to be the alpha male and dominate the rest, a status for which the males fiercely competed. After all, the alpha male gets all the young ladies and, being gourmands in things sexual, many of the rest. The conflicting half to our duality was that no man wanted to be subservient to another. It was our nature not to submit to being ruled; we resisted domination.

Males losing to the alpha male wound up with not only limited sexual access but also the humiliation of being subservient. Consequently, such societies existed in a state of chronic conflict, making cooperation difficult. The important point is that men in conflict would lose in war against a society of cooperators.[18] In war, internal conflict decreases fitness.

This is our dichotomous starting condition, one from which we became egalitarian. Once we did, there was no ruler and no one was ruled. How did we make this change?

It is my belief that the change we are about to discuss, the one to egalitarianism, is the start of becoming human. Before, we were simply just another, probably somewhat nasty and brutish hierarchal primate in Boehm's ancient forest. This is the beginning of our humanness, the genesis of man's modern psychological nature. It is also the start of humans becoming moral-making machines. We start to develop a better nature.

Becoming Egalitarian

Egalitarianism is not the simple "all men are equal"; it is more nuanced. Egalitarianism is both the absence of authority figures and the proactive insistence on the equality of all persons. It additionally includes the refusal to bow to authority.[19] This is the heart of Boehm's thesis and the topic for this section.

His thesis posits that we changed from our chimpanzee-like, hierarchal nature to an egalitarian one through two processes. The first was group repression of despotic behavior, which effectively eliminated the alpha male and its associated behavior from our midst. Repression of proscribed despotic behavior employed group-sanctioned punishment, ranging from shame and ridicule for mild offenses to expulsion and execution for more serious ones.

According to Boehm, there is a very important element to egalitarianism. That element is the proactive nature of repressing despotic behavior. The kernel of his thesis is "that egalitarianism does not result from the mere absence of hierarchy, as is commonly assumed. . . . It is based on antihierarchal feelings."[20] In other words, becoming egalitarian required proactive actions against any expression of hierarchal behavior.

The focus of his thesis is on despotic related behaviors. They include alpha male behaviors, such as arrogance, bragging, and over-assertiveness, and upstart behaviors. Though Boehm's focus is on despotism, his list includes stinginess. He emphasizes, "Stinginess is a serious form of deviance . . . , but it is the arrogant person who is potentially dangerous."[21]

How did the group minimize despotic and other antisocial behavior? According to Boehm, the group exercised a continuum of antiauthoritarian sanctions. They ranged "from moderate (criticism, ridicule, disobedience) to strong (ostracism, or expulsion, deposition or desertion) to ultimate of execution."[22]

Persistent or particularly odious behavior elicited expulsion or community-sanctioned execution. Our ancestors were very serious about maintaining equality and not letting anyone take charge.

Men who were disposed to despotic behavior *and* were insensitive to group sanctions were a singular risk to the group's survival. For them, the only effective recourse was probably group-sanctioned expulsion or execution.

Even though expulsion and execution are a fine means of ridding a gene pool of undesirable traits, the burden of insensitive and despotic men persists into our modern era. This persistence suggests that egalitarian societies repressed, but did not eliminate, unwanted behavior. Culture masked genetic expression. Because natural selection can select neither for nor against any behavior that is masked, such traits are maintained and not eliminated. It may not have been nice, but we did fool Mother Nature.

The second process for becoming egalitarian was exchanging the dominant authority of the alpha male with that of the collective. We, as the collective, became boss, replacing the role of the alpha male. However, we remained subservient. The difference was that we were now subservient to the collective, not some brutish alpha male. Additionally, being a voting member of the collective gave us some rights; we had a say in how we were bossed.

There is an alternative perspective. We replaced the tyranny of the individual with the tyranny of the majority. For better or for worse, this is what we did.

This fixed the problem of not wanting to be subservient to another individual. Innately despotic alpha males who simply could not stand living in a society with another as boss found this compromise acceptable. As long as the collective persisted in enforcing a politically egalitarian society, all, including alpha male wannabes, were reasonably content. For most, life, though not perfect, was acceptable.

Boehm provides the parsimonious explanation: "All men seek to rule, but if they cannot rule they prefer to be equal."[23] Those not willing to accept the dominant role of the collective had the option of leaving. However, in those primitive times, leaving the protection afforded by the group could easily result in death. To acquiesce to the demands of the group acting as the dominant authority was the prudent choice.

Egalitarianism was not a constant. The family is a unit of an egalitarian society, and behaviors between individuals of a family unit are somewhat outside the bounds of communal sanctions.[24] According to Boehm, "authority tends to be present, legitimate, and relatively unrestrained within the household. Dominant control is directed at children, and often at wives."[25] Behavior within the group that could warrant execution might be acceptable within the family. Double standards are not new.

Egalitarianism became the acceptable compromise; no one could be ruler, no one had to bow to another, and everyone had to accept the rule of the majority. Egalitarianism was king.

Culture: The Enabling Agent for Egalitarianism

The thesis we discussed about the development of egalitarianism is not complete; interesting questions remain. There has to be more than the unadorned "we repressed despotic behavior."

Why is it that, of all the primates, only humans developed egalitarian societies? Assuming that egalitarianism is adaptive, we might expect that some other primates would have become egalitarian, yet they have not.

There is an unstated premise to this assumptive statement: it is that our change to egalitarianism was due to conventional, genetically based evolution. Perhaps it was not our genes that were responsible but rather our humanly unique culture. Perhaps we would best view our genetic first cousins, the chimpanzees, as our cultural fourth cousins twice removed. The obvious point is that the cultural gap far, far exceeds the genetic.

As humans, we have culture, but this has not always been the case. At some time, probably while we were still in Boehm's ancient forest, we lacked significant capacity for culture. In this aspect, we were then more similar to our chimpanzee cousins than we are now. Between then and now, our capacity for culture evolved. If culture is the mechanism enabling egalitarianism, the answer to why we, and not our cousins, made the step is clear. We could, they could not, and most importantly, it was adaptive.

There is another argument against conventional, gene-based evolution being the mechanism for changing our social nature from hierarchal to egalitarian. If it were based on such evolution, the process would have had to beat back innate despotic tendencies concurrently with the development of egalitarian ones. This would probably have been a relatively slow process, for as soon as the egalitarians had gotten an edge, the despots would have found ways to recover and remain dominant. It would have been an ineffective, whack-a-mole strategy, and an egalitarian society probably could not have developed.

There are other aspects to this question. The movement from despotism to egalitarianism appears to need an all-or-nothing event, something like a step function or saltation. Genes do not step or jump; their allelic frequencies change smoothly over time. Genetic evolution takes multiple generations; culture can be quicker to adapt. It is easy to fashion a scenario where human groups, functioning in a cooperative manner, use culture to shut down despotic behavior. A like-minded group could start in an afternoon. All it takes is group agreement, firm commitment, and faithful execution of a plan over an extended period.

The cultural, proactive repression of despotic behavior did not modify its genetic basis; it only eliminated the effect of the selfish, dominant male on society. The genotype remained despotic, but the phenotype became egalitarian. We hid our true nature from Mother Nature.

Without the presence of dominant males, other males could acquire mates, resulting in increased group cohesiveness. Assuming that cooperative groups outcompete selfish ones, this could represent a form of positive evolutionary feedback for reinforcing cooperative behavior.

War Promoted Egalitarianism

Evolution is about competing and winning. When individual selection is in play, individuals compete, and likewise, so do groups in group selection. Before we repressed despotic behavior, men fought, with proprietary sexual access going to the alpha male winner. With group selection, the battlefield was the field of competition. Winners lived to fight another day. Losers didn't.

An important insight into the development of human nature is in the following quotation from Edward O. Wilson. It explains the evolutionary consequences of societies with altruists:

> Competing is intense among humans, and within a group, selfish individuals always win. But in contests between groups, groups of altruists always beat groups of selfish individuals.[26]

Groups of selfish individuals are easy to find. Just look at any social group in our modern times. Groups of altruists are scarcer.

In the following, we are going to show the connection between egalitarian and altruistic groups. In a condensed form, the story is as follows. We became egalitarian by suppressing despotic behavior. This decreased individual variation, enabling group selection, a consequence of which is the superorganism. Furthermore, there is a requirement for the individuals of a superorganism to behave altruistically. So, what we have are groups that are egalitarian and with group selection. They are superorganismic and composed of altruists. For human societies, they are all of a piece. Developing an understanding of this aspect of our nature is what the following is about.

On the basis of the foregoing quotation, if there are two societies, one of which suppressed despotic, alpha male behavior and the other of which did not, the first would dominate the second in any conflict. This means that repressing despotic behavior increased group fitness by increasing internal cooperation.

The preceding argument is true only if war was an evolutionary force. If our nomadic ancestors had been peace loving, seldom engaging in war, there may not have been an evolutionary advantage to repressing despotic behavior and becoming egalitarian. We would have to find another reason.

There is an ongoing debate about the nature of man as it relates to war and violence. Is war an innate part of our nature? Were we peaceful denizens of this earth until we settled with agriculture or became possessive with cattle? There is a multitude of arguments, pro and con, for almost any stance. Some even have a rational, scientific basis.

Evidence going back to our starting date of a million years before the present is nonexistent, so we will probably never absolutely know. However, evidence going back twelve thousand years[27] or so suggests that conflict between groups was a constant. If war was as prevalent two hundred thousand years ago as it was twelve thousand years ago, then it has been more than sufficiently prevalent.

Modern societies are not egalitarian. Does this mean that the evolutionary pressures for egalitarianism have disappeared? If the primary pressure for becoming egalitarian had its roots in human conflict, the answer is yes. Modern man is far less violent than were our nomadic, foraging ancestors.[28]

This is all to suggest that natural selection at the group level resulted in selection of groups repressing despotic behavior. Groups not doing so and thereby not taking up the banner of egalitarianism fell by the evolutionary wayside. The winning egalitarians literally became the chosen ones.

The conclusion to this part of our story is that evolutionarily successful groups were those who reduced internal conflict created by the despotic behavior of individual males. Using unity of action, the group applied group-sanctioned punishment to repress the despotic behavior of individuals, resulting in groups having increased internal cooperation. Such societies were better at competing in war. Consequently, in response to the evolutionary pressures of intergroup conflict, we became egalitarian.

Repressing Despots Unlocked the
Power of Sexual Selection

Until we repressed despotic males, we were doomed to being just another ape in the forest. Sure, natural selection would have worked to keep us surviving, but sexual selection was locked into a singular mode, that of the dominant alpha male. A persistent condition of sexual selection based on the alpha male would have kept our psychology frozen in place. We could never have changed our ways. We could not have advanced.

Sexual selection is a mode of natural selection with the following definition:

> Darwin identified two separate (but potentially related) causal processes by which sexual selection occurs. The first, intrasexual or same-sex competition, involves members of one sex competing with each other in various contests, physical or otherwise, the winners of which gain preferential sexual access to mates. Qualities that lead to success evolve. Those linked to failure bite the evolutionary dust. Evolution, change over time, occurs as a consequence of the process of intrasexual competition. The second, intersexual selection, deals with preferential mate choice. If members of one sex exhibit a consensus about qualities desired in mates, and those qualities are partially heritable, then those of the opposite sex possessing the desired qualities have a mating advantage. They get preferentially chosen. Those lacking desired mating qualities get shunned, banished, and remain mateless (or must settle for low quality mates). Evolutionary change over time occurs as a consequence of an increase in frequency of desired traits and a decrease in frequency of disfavored traits.[29]

A classic example of sexual selection is the peacock's tail. In mate selection, the peahen selects the peacock with the best tail display. From the peacock's perspective, it competes for sexual access with tail displays. This specific case of sexual selection ever pushes the genes of peacocks toward grander, peahen-appreciated displays.

A key consequence of sexual selection is that

> a trait that makes an individual more likely to be selected as a mate will provide its owner with considerable reproductive advantage, and, in consequence, it will increase in frequency in the population.[30]

As long as humans had despotic males dominating the reproductive process, the despots' genes were the ones prevailing in the population. Genes for *unokai* warriors, *tokwaybagula* gardeners, or rich English merchants could never flourish; they were not even in the game. It was with repression of despotic behavior that these and other traits could become the basis for sexual selection. We, or rather our populations, could change, and change we did. Depending on our respective evolutionary paths, we became better warriors, gardeners, or merchants.

Egalitarianism Decreased Variance, Thus Enabling Group Selection

On the basis of the ethnographic observations presented in the first half of this chapter, we made the case for human group selection and the superorganism. In this section, we will see how egalitarianism enabled that process.

The following quotation clearly defines the problem:

> The key question remaining in the dynamics of human genetic evolution is whether natural selection at the group level has been strong enough to overcome the powerful force of natural selection at the level of the individual.[31]

Independent of the type of natural selection, group or individual, selection pressure depends on the size of variation of the trait under selection. Greater variation results in higher pressure. Normally, the variation between individuals is greater than the variation between groups. Therefore, for as long as this condition holds true, group selection should not occur.

The foregoing is a main argument against group selection. Clearly, for a given trait, if individuals within groups were equal but the groups themselves were not, group selection for that trait could occur.

Egalitarianism tended to make individuals equal, or at least to appear so to the forces of natural selection. Equalization resulted in reduced variation between individuals, and zero variation would be the consequence of perfect equality. The degree of egalitarianism we achieved was apparently sufficient in reducing variation between individuals to less than between groups. This is what enabled group selection.

Recall from an earlier section that natural selection operates on the phenotype, not the genotype. Even in the presence of genetic variance between individuals, if culture masks those differences, the phenotype can be the same. Natural selection has no basis for choosing if the phenotypes are equal, and in that situation, individual selection cannot occur. The formation of egalitarian societies created such a condition. With sufficient reduction of between-individual variation, group selection was enabled and our kinship groups became superorganisms.

Enabling group selection did not eliminate individual selection. It still operated, but now within the context of the group and group selection. The superorganism, as a survival machine, faced the outside world and had to meet its demands for survival. All the individual had to do was contend with the environment within the superorganism, especially its internal demands. The system of the superorganism was such that if the individuals did their jobs, the system in turn met their existential needs. Survival was a function of the superorganism, while serving the superorganism was a responsibility of the individual. The individual worked for the superorganism, not self, a defining characteristic of the human superorganism.

Being a member of the superorganismic, group survival machine is not some casual relationship; it is all encompassing and all consuming. It is something for which we are willing to die. It, and not the self, is our interface with the world. Our survival is dependent not on self but on the survival of our group survival machine. We all survive, or no one does.

Most of the literature in evolutionary biology is concerned with general rules that apply to all the world's organisms. The arguments we have made for group selection are atypical in relation to the entirety of biology because we humans, with our culture, are atypical. We are unique in the degree of our culture; no other organism even comes close. The culture of human egalitarian society enabled group selection and the changes that followed.

We started this chapter discussing the debate concerning group selection as a viable process. If there were no culture, especially one that could mask inequality, group selection based on genetic variation could probably not be realized. It would not be viable. However, we have culture with the ability to mask inequalities between phenotypes, thus enabling group selection. What we have is group selection founded on culture, not genes.

If we had group selection with a genetic basis, we would always form such groups; it would be our innate nature. We would be just like the social insects with their deterministic social organizations. That we do not suggests otherwise. This is not a case of cultural group selection, where selection operates on the cultural differences between cultures. This is group selection enabled by culture—a different process.

We are now in a time before the Urban Revolution, a time when we live in superorganismic groups. Additionally, these groups are the unit of natural selection and individuals within are altruistic. We got to this place because we lived in an environment where groups competed one against the other in war, making egalitarianism, with its consequential altruistic behavior, adaptive. Remember, such groups beat those not having this trait. The consequence was that egalitarianism led to group selection, the superorganism, and altruistic individuals.

CHAPTER 8

Nature of the Superorganism

In our model, we now have pre-state societies existing as superorganisms. When some arrived at the portal of Urban Revolution and the formation of the state, not all entered. They stayed behind and remained psychologically as primitive man. They are today's poor. Today's rich are those who made the transition, underwent changes, and became psychologically modern man. Some societies are still transitioning and exist in an in-between state; they, too, are typically poor.

Independent of their status, at one time all societies were superorganisms. Given this ubiquity, it is time to put more clothes on this idea of a human superorganism. That is what we are going to do in this chapter.

The first part is a comparative analysis between superorganismic bees and human psychology. It is a discussion of the characteristics of the generic superorganism and the correlating properties of human psychology. Here we will examine the characteristics of human psychology as it applies to being a superorganism. That will be a view of the group. Later, we will view the nature of the individuals within.

Humans and Bees: A Comparative Analysis

Introduction. Superorganisms, just as organisms in general, have common characteristics. Modern man might not live as superorganisms, but our human psychology maps very nicely onto that of the generic one. The fact that our psychology has much in common with the hypothetical superorganism is consistent with the thesis of our societies at one time being so.

Aiding our discussion is a paper by Kesebir that makes such a comparison.[1] In it, she compares our psychology to that of the superorganism as exemplified by the bee. Kesebir's comparison is between modern human societies and bee societies. Our concern at this time is with early, nonmodern man. To fit our needs, I have taken her base concepts and transposed them to what I believe was our earlier nature.

There is an interesting conclusion to her paper. Modern man has many properties of the social insects, yet we do not live in communal hives or mounds. We must have acquired this nature somewhere, and our story of why we became superorganismic ties into that analysis. Furthermore, what it suggests is that traits, once acquired, can persist into the future, well after their originating purpose is well past. In many cases, we are who we were some very long time ago.

In making the comparison, Kesebir uses the following five characteristics. We will use the same:

1. integration of lower-level units through communication

2. unity of action

3. low levels of heritable variation among the superorganism's units

4. a common fate

5. mechanisms to resolve conflicts of interest in favor of the collective

Without further ado, we will examine these traits as applied to our psychology.

Integration of lower-level units through communication. For the units of a superorganism to function as a group, they must be able to communicate. However, this does not necessarily mean language. Ants do quite well with chemicals and bees use body language (wing wags). Humans have superb communication abilities of various types, and when we choose, we can communicate very well.

We can speak, and many can write their languages. Additionally, when in a strange land, we can draw pictures of what we want. The other can similarly respond; both are understood. However, we do not require explicit language because, like bees, we have nonverbal language. Examples are blushing, stammering, and looking mad or happy. When we converse, we look others in the face to receive the entirety of their communication, not just their spoken words. Not only do we make nonverbal expressions, many involuntarily, but also when expressed by others, we unambiguously understand them.

There is an interesting question associated with this property: did our language coevolve with our becoming superorganismic? Language is adaptive in and of itself, and for certain environments, so is the superorganism. It is straightforward to consider human evolution advancing toward superorganismic groups hand in glove with the development of language, both nonverbal and verbal, the development of each reinforcing the other.

Unity of action–shared intentionality. Superorganisms have the capacity to function as a unit, and this translates into shared intentionality. Our intelligence and communication skills enable our ability to collaborate. Three examples in modern man are the New Orleans Saints football team, Microsoft, and the construction crew building the condominium down the street. These are all groups whose several individuals know what the intended group goal is and work toward it, be it more touchdowns for us and fewer for them, a higher stock valuation, or a building properly constructed to plan.

Even though the foregoing examples reflect the property of the superorganism, they are not superorganisms. Their behavior is probably a remnant of a superorganismic past.

To examine this characteristic within the context of the superorganism, we need only look at the Yanomamö. When their men go hunting, they do so as a cohesive group with shared intentionality, not as a bunch of individuals. The same property is evident when the man and women coordinate their labor to prepare a virgin forest area for a garden. Men chop down trees, women tend plants, and children pull weeds. In both hunting and garden preparation, everyone knows the desired result, and each knows all the collective steps required. They all know their expected roles.

Modern man exhibits the same type of shared intentionality as primitive societies do. This suggests that behavior such as that exhibited with shared intentionality, once learned, persists into the future.

Unity of action–social identity. The discipline of psychology uses Social Identity Theory to explain group behavior. This theory posits that group processes are more than the simple sum of individual ones. Social identity creates a whole that is greater than the sum of its parts.

We have discussed that a characteristic of the Urban Revolution was changing identity to the state from the prior kinship group. That discussion was about social identity.

An example of social identity is when group membership informs our sense of identity. Belonging to a group is more than the objective membership in a neighborhood organization; it is subjective. Because of social identity, we incorporate the group into our concept of self; this is key. There is an emotional component; when our team wins or our group succeeds, we feel good.

There is a spectrum of social identity intensities. It can be as simple and low key as feeling good about wearing the colors of your favorite sports team or green on Saint Patrick's Day, provided you are Irish. At the other end is a willingness to commit the ultimate sacrifice for the group. Belonging to a group with a high level of internal social identity brings a strong sense of "all for one and one for all." We are the group and the group is us. We belong to the in-group, others to the out-group. We are good; they are bad.

Social identity has other properties. It leads us to give preferential treatment to in-group members; we are loyal to our group's members; we have ownership of the group's goals. Importantly, as our identification with the group intensifies, we move away from our perception of self as a unique individual. Quoting Kesebir,

> the incorporation of the group into the self-concept marshals the same cognitive, affective, and behavioral processes associated with the self. Cognitively, groups become included within people's mental representation of themselves.[2]

A consequence of social identity is placing others into groups; we classify others. We identify with our group(s) and, wrongly or not, assign others to groups. We categorize others as we do ourselves. Others belong to an out-group and we are the in-group. Would we discriminate against the out-group? Of course we would, and we do.

Social identity has a role in our self-esteem, which can be raised by promoting our group(s). This is a zero-sum game where we promote our group and thereby ourselves when we denigrate the out-groups. We go up and they go down. When the Saints play the Panthers, Panther fans are the enemy. To a Saint fan, they are lower than low, at least for game day. What do the English think of the French? Never mind asking the French.

Unity of action–deference to legitimate authority. This is a characteristic Kesebir has right, but for the wrong reason. Individuals of the superorganism do defer to legitimate authority, but her assertion that legitimate authority resides with high-status individuals is not quite correct. The superorganism is egalitarian, and in such societies, everyone has the same status; there is no high-status individual. However, that does not mean that there is not a legitimate authority. There is; it just happens that it is the superorganism itself. It seems that having a boss is a requirement for any society, and because egalitarian societies proscribe authority figures, the task of being boss must fall to society itself.

Before making the transition of the Urban Revolution, we had a long history of conforming to the desires of the superorganism, which we did whenever we deferred to its legitimate authority. With the Urban Revolution, the superorganism as a legitimate authority no longer existed. Nevertheless, our nature maintained its responsiveness. We did not change; our bosses did. The new ones were now individuals, not the collective. This is another example of past behavior carrying over to a new, different, modern environment.

In modern, nonsuperorganismic groups, such as the state, the military, and the company we work for, we continue to defer to a presumed legitimate authority. It just is no longer the superorganism.

Unity of action–self-organization. Self-organization into a superorganism is one characteristic that humans do not have in the same sense as other superorganisms. Bees and other social insects simply self-organize. Take some bees, place them in a metaphorical bag, mix them up, and after a time, bee hives will appear. Do the same to humans and there is no self-organizing into a superorganism. We are soft-wired. For us to get to that type of organization, we require prompting from our environment.

The environmental prompt appears to be a need to organize in order to provide for a common defense against between-group conflict such as war. The first rule of evolution is survival, because getting to the reproduction part requires survival. As we have discussed, the environmental prompt for getting to our initial superorganismic state was war. Societies that repressed despotic behavior, became cooperative egalitarians, and evolved into superorganisms outcompeted in the field of war those who did not. They survived; the others did not. Warriors who cooperate beat warriors who do not.

In today's modern world, the state provides protection from exterior forces; we do not have to defend ourselves. If the state were to disappear, like in some apocalyptic movie, would we again resort to a superorganismic state? We might, provided we chose to become egalitarian, as that is the requisite route to being a superorganism. If we remained hierarchal, as we are now, the answer is no. Can we be hierarchal and still provide for our common defense? I see no reason why not.

Modern man was once superorganismic, but that was in the past. He is not now, and returning to such a state would be problematic. The important point is that behavior that evolved in our superorganismic past persists today.

Low levels of heritable variation among the superorganism's units. Colonies of social insects, like bees and termites, are classic examples of the superorganism. They have a low level of heritable variation between individuals, which is due to the way they reproduce. Essentially all the individuals in a hive or colony come from a single queen. This results in each colony having a high level of genetic relatedness and a concomitant low level of heritable variation between individuals.

This is fine for the bees and termites but has nothing to offer humans. A human group with just a single breeding female would be interesting indeed. We had to use a different path to achieve this aspect of our superorganismic nature. This is what we did in using our unique culture for creating group norms. The group norm for egalitarian behavior resulted in effective equality between individuals, translating into low expressed variation.

Using culture to create low variation is a double-edged sword. It is simple to do and takes little time. At the same time, it lacks the permanence and deterministic nature of genes. It can be vanquished as quickly as it is acquired. Importantly, it only masks genetic proclivities, which persist, ever ready to re-present themselves to natural selection once the lid of culture is removed.

Just because culture lacks the deterministic nature of genes does not mean it lacks heritability. Inheritance of culture is via learning and occurs when it is passed from parent to child or from a group of old-timers to recent immigrants. On joining a new group, we are conditioned to observe and emulate the others. After all, if we want to belong, we should be the same as, not different from, the group to which we wish to belong.

A low level of heritable variation between individuals is not only a characteristic of superorganisms. It is also a requirement for its formation as well as another property of the human superorganism: group selection. Suppression of traits exhibiting dominant behavior along with proactive fostering of egalitarian ones enabled the human superorganism as well as group selection.

Owing to two changes associated with the Urban Revolution, we lose the property of low heritable variation. First, our societies change from egalitarian to hierarchal. Hierarchal societies, in contrast to egalitarian ones, permit individual differences, thus increasing variation. Second, the individual changes identity from the superorganism to the state. This enables individualism, further amplifying the change from egalitarian societies. Living in states as individuals and citizens, modern man is not superorganismic and is again the target of natural selection.

A common fate. Kesebir has common fate as a property of the superorganism. On this she is correct, but the perspective of her examples is misleading, as they take the view of the individual and not of the group. The group view is where the importance lays.

The basis of one part of her discussion is egalitarianism. Here she correctly argues that a consequence of the cultural norms of an egalitarian society is equal outcomes or a common fate. Another part of her discussion has intergroup warfare as its focus. War's consequences, especially for losers, usually result in an unpleasant common fate, but a common one nevertheless. Both of these points play to the view that common fate is all about the group, not the individual.

The importance of common fate is the fact that it aligns the interests within the group. What happens to one happens to all; it is the classic Three Musketeers, all for one and one for all. It is the egalitarian ethos where no one starves unless all starve. This view from the group, and not its individuals, is existential, as it is literally life or death. If the superorganism dies, so do its individuals. Superorganisms are simply another class of organisms, and when an organism dies, so too do its cells. This view of the superorganism is one as a survival machine and one that includes group selection.

Of all the properties of the superorganism, this is one that is not reflected in modern man. Granted, modern man can and does form groups where individuals have a common fate. Football teams and companies are but two examples. However, neither comes close to the absolutist nature of life or death encased in the group as a superorganism.

Mechanisms to resolve conflicts of interest in favor of the collective. This characteristic is different from the others we have discussed; it applies only to humans, not bees or other social insects.

A problem for the human superorganism is cheaters or free riders. Such miscreants reap the rewards of the public good provided by others but offer nothing in return. If their behavior is unchecked, everyone will become a cheater, as there is zero fitness disadvantage in doing so and every advantage in being one. Unless cheaters are suppressed, everyone becomes one, and society ceases to be an "all for one and one for all," cooperating superorganism.

Much of our group psychology has evolved for solving this problem. Just as we suppressed dominant male behavior to become egalitarian, we use group-approved sanctions to punish cheaters. These can range from simple group disapproval to group exclusion. Importantly, to be effective, individuals must be sensitive to the disapproval, and it turns out that we are averse to such criticism and ridicule. We have evolved to be sensitive to group norms and to group responses when we transgress those norms. Evolution has covered all the bases by providing both approbation for good behavior and disapproval for bad. Individuals seek the first and behave in a manner to avoid the second.

Psychopaths lack sensitivity to group approval or criticism, meaning that they cannot be controlled through normal means, making them an especial danger to society. In such cases, groups must use extreme mechanisms, such as execution, to effect control.

Being sensitive to group approval or criticism is a trait just like the others we have discussed. However, this trait appears to be a requirement for the human superorganism. An important point is that, like the others, it persists in the psychology of modern man.

Kesebir ties this part up nicely:

> In sum, human psychology is configured to make cooperation with group members rewarding and non-cooperation distressing. In addition, cultural practices that build on capabilities such as language and normative compliance serve to align individual interests with collective interests. As a result, cooperation is often a win–win strategy for members of a group, whereas selfishness is often self-defeating.[3]

Social Control: A Property of the Human Superorganism

Introduction. Our ancestor societies, the superorganisms of the past, were social organizations that evolved methods and means for directing and controlling the behavior of their individuals. They learned that the collective, acting in unity and using group-mandated punishment as a tool, could prevent unwanted behavior. Repressing despotic males was probably the first instance. Others, like punishing thieves and liars, would follow.

Repressing unwanted behavior is only one side of the coin. Promotion of desired behavior is the other, and this they learned how to do by rewarding such behavior with praise. Our psychology evolved to seek praise, and we directed our behavior toward that end. We still do.

Praise is significantly less of a lever than punishment; it has nothing even close to the ultimate punishment: execution. However, the group did have in addition to praise and punishment another hammer in its tool box. We were clever and found a way to use a very big hammer indeed, management of Darwinian fitness. We learned that by managing the social rules for mating, we were able to reward those doing good for the group with increased fitness, an example being extra wives for the *unokai*. Moreover, we could do this at no cost to the group. This mechanism led to the evolution of altruistic behavior.

We are going to start with plain vanilla punishment and then move on to praise. We will then spend time on the interesting topic of using reciprocal fitness to promote the group-desired behavior of altruism.

Becoming egalitarian. We have used Boehm's thesis of becoming egalitarian to explain forming superorganismic groups. If we look at egalitarianism from the perspective of the individual, we see the obvious equality of individuals. The perspective from the group is more complex. It is also more important.

Our transition to egalitarian societies took place when we were chimpanzee-like. The group took charge, repressed despotic behaviors, and then proactively enforced egalitarian ones. With this step, our societies changed from hierarchal, with the alpha male being in charge, to egalitarian, with no single individual in charge; we all were.

An important point is that when we repressed despotic behavior using appropriate punishment, we learned that we could control other unwanted behavior. We did not have to suffer liars, cheats, and thieves. The first step was identifying and agreeing to the behavior the collective found objectionable. The next was to establish appropriate punishment, and the last was to administer it consistently. This was the first type of social control. In time, individuals learned not to behave in a manner the group found objectionable, as such behavior resulted in negative consequences.

This was a new management tool. We could decide how people ought to act and, in the case of miscreants, inflict punishment until they conformed. At one extreme, we could execute or expel for serious offenses like murder, and we could apply the same treatment to dangerous psychopaths who were indifferent to punishment. At the other end was shaming and similar group-exercised punishments. We, as the collective, got our way over behavior of the individual. We, the superorganism, were the dominate authority.

This is the single step in our evolutionary journey that is central to understanding who we are. With this initial tool of social control, we could mold social behavior and, in so doing, become moral animals. We changed to a society where we were all in charge and became, in the process, a collective moral-making machine. Culture could now direct traffic and put us on evolutionary pathways not otherwise available. Any group so constituted could take any path it wanted, and we started becoming different.

Because of our vast capacity for culture, only humans were able to make this change. We created societies where the collective decided and enforced the rules for social behavior and, with this, established morals. Our behavior evolved in this environment, and the process of becoming egalitarian marks when we started to become psychologically human. This became our ancestral home; this is where we grew up.

Consequences of social rule making. Getting to self-sustaining, egalitarian societies probably took time, as we had to develop numerous social rules to be effective. However, in the end, the superorganism is in charge, and it, through the agreement of the collective, makes the rules for all to live by. Additionally, it establishes penalties for transgression of those rules. This is a bottom-up, majority-rule collective, establishing the superorganism as the dominant authority.

Within this scenario are two parts; one is the group and the other comprises the individuals within it. Each has its respective role. We, the you and I composing the group, make the rules, and as individuals, we obey them.

The superorganism, as the dominant authority, has absolute power over the individual. Its power ranges from execution or, nominally equally fatal, expulsion for egregious despotic behavior to rewarding with increased fitness for altruistic behavior toward the group, a topic we will discuss later.

The nature and exercise of that power is dependent on agreement within the collective. The important point is that it is not the whim of some alpha male despot. The collective is the moderator of the superorganism, whereas the despot has no such power.

This is a majority-rule, bottom-up system where the result is social behavior desired by the collective. If the collective wants equal sharing of the spoils of the hunt, then they are. Otherwise, the collective metes out agreed-to punishments to those not sharing. The social behavior aspect of culture is the agreed-to behavior of the collective.

This creates an interesting entity. Individuals follow the dictates of the collective, and all the while, the individuals are the collective. The nature of this relationship makes circumspection the prudent mode and is a force for moderating any tyrannical tendencies by the group. You never know if you are going to be the collective goose or the individual gander. Since what is good for one is good for the other, behave and decide accordingly!

It is my view that this is where we break with the rest of the animal kingdom and change from being just another primate in the forest. This is where culture enters the scene and, at the same time, begins to develop our humanity. In addition, this is where we start to become what we want to be.

With punishment as its sole initial tool for social management, the collective could minimize unwanted behavior. Our tool kit for social management probably started with only the tools of expulsion or execution, with their use being restricted to serious offenses such as consistent bragging or bullying. In time, these tools expanded to include less onerous punishments, such as shaming, shunning, and ridicule. With these, we could proportionally address less serious offenses, such as lack of respect to an individual or not repaying a neighbor for help.

Desired behavior. No social animal enjoys group disfavor, making punishment an uncomplicated and useful stick for preventing unwanted behavior. However, it is a poor one for promoting wanted behavior; for that, we needed a carrot. Somewhere along our evolutionary journey, group rewards for desired behavior became a social management tool. For this, the group evolved to give praise and granted rewards for desired behavior.

Why did the use of group rewards evolve? It probably did for the same reasons underlying other evolutionary outcomes. Societies rewarding wanted behavior outcompeted those that did not. Remember, with the superorganism, group selection is the *modus operandi*.

For group rewards to be effective and result in desired behavior, two parallel steps were required. The first was that, just like unwanted behavior, the group had to decide what it wanted. The second is that individuals had to evolve psychologies for seeking group praise. They did, and we do. Our social group giving us praise for our actions makes us feel good, and we seek to repeat the experience.

There is a general form to the foregoing—status. For a variety of behaviors, we acquire status from the group. This can come from helping an old lady cross the street or defeating an existential enemy in mortal combat. Importantly, each behavior comes with a reward proportional to its resultant value to the group, with greater value resulting in higher status.

Status of this type seems to be associated with reproductive fitness. For example, high-status males have more mating opportunities—consider the *unokai*.

However, there is more. In mate selection, fitness indicators, such as large antlers in deer or youthful appearance in humans, play a role in evolution. Those possessing these indicators have better mating opportunities because they represent a greater probability of their genes going forward. Proactively seeking a fit mate is an evolutionarily sound strategy.

Status of the sort we are discussing is a symbol of fitness. Heritable traits leading to the reward of status, just like big antlers, go forward. Populations change accordingly. They change to reflect the behaviors preferred by the superorganism, and they become what they want to be.

Group preferences leading to group-rewarded status, in turn leading to reproductive fitness, is a powerful mechanism. In a like manner, there is also decreased reproductive fitness associated with behavior disapproved of by the group. If there is a single message in our narrative as to why people are different, this is it. Importantly, the driver is culture.

We now have a complete system, one that is majority rule and bottom up. The group decides what behavior it likes and dislikes. It has tools to prevent one and promote the other. This is an evolutionarily powerful machine. Our ability for social control is what gives us culture and morals, traits not found in other animals.

Morals are adaptive. When we agreed on prescriptive and proscriptive behavior, we were agreeing to morality. We decided the "ought to" of life. Rightly or wrongly, we were the deciders of right from wrong. We were the moral-making machine.

When we started, there was no religion or philosophy guiding us, just evolution. Morals are specific to the ethnic group creating them, and there is no *a priori* reason for any two groups to have the same morals. *A posteriori*, there is.

Morals are what individuals have agreed to about how to behave, especially with regard to how to treat each other. They are a group property, and therefore, in group selection, they are a selectable trait. In the competitive evolutionary world of group selection, groups having the "best" morals win. The process of evolution decides what is "best," not some elitist philosopher in an ivory tower or crazed monk in a hair shirt. Independently of any modern-day relativist's paradigm, there is and always has been an absolutist aspect to evolution. Independently of any social need for truth in cultural relativism, biology is absolute. Only the genes of the culturally correct propagate. Mother Nature *does* know best.

If populations have similar environments, "best" should have some level of concordance, and there should be some level of commonality in morality between groups. Additionally, morals are a meme of culture, meaning that groups can learn and assume the morals of others. Consequently, with both processes, we would expect a large degree of similarity of morality between groups.

Altruism: A Product of Social Control

Introduction. The prior section was about social control. It was about the superorganism preventing unwanted behavior and promoting wanted behavior. In this section, we are going to see that a form of altruism exists because of social control. It is a form of behavior actively promoted by the superorganism, and what appears to be altruistic behavior is, in actuality, not.

Labor for the superorganism. We earlier established a metric for determining if a group is a superorganism. It is in the answer to the question, who does the individual work for, self or group? If the individual works primarily for the good of the group, it is a superorganism.

In the ethnographic observations we examined, the motive for labor was to gain social reward or status. This is what competing to gain the title of *tokwaybagula* was all about. Laboring for the group and gaining its reward of status has all the appearances of altruistic behavior, as status certainly does not fill the belly. However, appearances can be deceptive, both in life and in evolution, and what was described might not have actually been altruistic behavior.

Definition of altruism. There are two flavors of altruism, one moral, the other biological. From the perspective of an evolutionary biologist, altruism decreases the fitness of the altruist while increasing that of the recipient. Intent, as in the moral perspective, is not important. Only consequences for reproductive fitness are, as, for our analysis, this is a biological and not a moral issue.

Working for the group and thereby increasing its fitness is usually a straightforward proposition and unambiguously meets the first part of the definition. The second half, where the altruist loses fitness, is often not as clear-cut. As we work our way through the issues, we will see that the question of whether there is a corresponding loss of fitness is critical.

Altruism and the superorganism. The gardeners of the Trobriand Islanders, which we discussed earlier, provide a good example of altruism and the superorganism. The fruit of their labor, yams, goes to others in their community and not to the actual growers. The only possible reward for the yam growers is gaining the title *tokwaybagula,* which is for exceptional gardening and gives the titleholder status. For the majority, the so-so gardeners, there is no such reward, and their yams *still* go to the community. They get nothing in return for their work, and for this group, labor appears to be purely altruistic.

To meet the definition for altruistic behavior, the gardener must lose fitness, and because he gets nothing for his labor, it appears as if the so-so gardener does lose fitness. However, the tribe's social rules for food distribution ensure that all are fed, where "all" includes the so-so gardeners. This suggests that the so-so gardener does not absolutely lose fitness. If he does not work very hard but still receives food, there could actually be a fitness increase. What seems clear is that in the overall system, there is neither a net increase nor a net decrease in fitness; it all appears to balance out. If the aggregate is the determining metric, then this is not altruism.

This apparent, but not actual, altruistic behavior is a characteristic of the superorganism. To function as an "all for one and one for all" system, this behavior is required, and it only looks as if the superorganism is populated with altruistic individuals. It is akin to a system where no one cuts his own lawn, just someone else's. In the end, all the lawns are cut, and assuming that everyone cuts a lawn, no one's fitness is increased or decreased.

Altruism increases group fitness. We used the following in developing our argument for the evolution of superorganismic societies. It provides evidence that altruism increases group fitness, as evidenced in a repeat quotation from Wilson. He states, "Competing is intense among humans, and within a group, selfish individuals always win. But in contests between groups, groups of altruists always beat groups of selfish individuals."[4] Adding to the evidence is a salient quote by Boehm: "A united moral community is a fearsome adversary indeed."[5]

As we have discussed, pre-state societies existed in a state of chronic conflict. If altruistic behavior resulted in the superorganism being better in intergroup conflict, it is fitness enhancing for the superorganism. As long as that condition were to persist, promotion of altruistic behavior as an element of social control would be predicted for such societies. It would be in the evolutionary best interest of the superorganism to promote altruistic behavior, and it did.

Sexual access. We have discussed that praise, although nice, is not as strong an inducement to directing behavior as punishment. However, increased reproductive fitness by increased sexual access is a strong inducement. This is the big hammer mentioned earlier. Access to women has been a major subtheme of our narrative. It is time to look more deeply and see how society uses it as the proverbial carrot in social control.

Like the social insects, the human superorganism has classes. Bees have workers—drones—and queens, whereas we humans have males and females. The female class, with a finite number of eggs, owns the factor limiting reproduction. The male class, with an essentially infinite amount of sperm, is unlimited in its reproductive potential. Controlling the limiting factor, access to eggs, is where the advantage lies.

Controlling the limiting resource for reproduction is power, but owning is different from controlling. Even though females physically own their eggs, historically they have not had absolute control. Although they are important players, there are two others with a vested interest. One is society. If it has control, it can use it to manage male access to women and, consequently, effect social control on male behavior.

The other players are men, specifically husbands, brothers, and fathers. Depending on culture, they all have had varying roles in controlling women and their limiting reproductive resource; just consider the symbolic act of giving away the bride in a Western-style wedding or a marriage arrangement in China. In a gene-centric world, women affect the destiny of the genes of their fathers, brothers, and sons; the latter are not disinterested observers.

For this part of our analysis, social control by society is our concern, and we can look to the Yanomamö *unokai* as our example. The *unokai* represents altruistic behavior. After all, what is more altruistic than going to war and perhaps dying for your tribe?

The social reward of *unokai* results in the successful warrior having extra wives. Does the extra wife have any say in being one? It doesn't matter. All that does is society and its culture having sufficient sway in the matter for effecting the important result: reproductive advantage for the *unokai*. The extra wife is simply a means to the end, as access to her eggs is the actual reward.

The Yanomamö, like countless tribes before, have a need to defend themselves against other similar tribes. If they are not successful in this endeavor, they could become history. To prevent this unwanted conclusion, their culture increases access to women to men who are superior warriors. Having such an incentive, their warriors should defeat others not similarly incentivized. The fact that the Yanomamö exist suggests the efficacy of this approach.

Establishing rules for mating gives society influence over an individual's reproductive fitness. In the foregoing example, *unokais* have three times as many progeny as non-*unokais,* a significant fitness advantage. For any young man, receiving as a reward for desired behavior an extra bedmate is a far greater incentive for behavior than the mere promise of an "atta boy" and the proverbial pat on the back. Deeds such as this speak far louder than any words.

The key issue is that societies, through their social rules, influence sexual access. This is a powerful tool for effecting group-desired ends, and society, by managing mating rules to achieve such ends, is using "free money." Males and females are going to mate anyway; all the group is doing is managing the process. Promoting behavior with mating rules has no cost to society.

Even though there is no cost to the group, the payoff potential is huge. Society, by staking out a management role in the sexual access game, is deciding what fit and not fit behavior is. Fit behavior is any promoted by the group resulting in increased access. The consequence is that any genes or heritable culture promoting this behavior will increase in the population.

The act of the group rewarding male altruists with greater sexual access might be the roots of a modern-day scenario. Similar to the *unokai*, the Islamic terrorist commits acts ostensibly for the good of the *ummah*, the collective community of Islamic people, by killing the enemy: non-Muslims. These two acts are different, because, whereas the *unokai* survives, the terrorist does not; the latter commits the ultimate, fitness-decreasing suicidal act. Somehow, Islamic terrorists are convinced that seventy-two virgins in the next life are adequate compensation for untimely death in this one.

Sexual asymmetry. In the preceding scenario, there does not appear to be a positive role for the female. The discussion about access to women was as if they were only sex objects; from the perspective discussed, they were—a consequence of both biology and egalitarian societies. We need to look more deeply to understand why this gender disparity exists.

Owing to biology, there is asymmetry in the reproductive strategies of human males and females. Each wants the same result: his or her respective genes propagated into the future. However, they go about it in different ways. Males, with unlimited sperm, commonly seek to impregnate as many women as possible. Quantity is their driver; they maximize their own fitness by siring as many children as possible.

Females, with a limited number of eggs, require a different strategy. They can have only a limited number of children and thus normally seek mates providing the greatest probability of ensuring survival of their children. The reproductive driver for women is quality; a woman's strategy is to maximize the fitness of her children, their survival, and not hers. In this respect, she is unselfish, in contrast to the selfish male interested only in his fitness. From a gene-centric point of view, a male or female strategy makes no difference; each results in the gene's propagation.

The process of altruistic men receiving increased sexual access is one consistent with the male reproductive strategy. Additionally, the altruistic act increases the group's fitness, making it evolutionarily sound from the perspective of the group. With this process, the group's fitness increases, and it, in turn, compensates the male altruist with fitness-increasing access. The obvious question is, why can't the same apply to women?

Awarding an altruistic woman some extra bedmates makes no biological sense. Having extra sperm to fertilize her limited eggs does nothing for getting her genes down the road; there is no fitness advantage. However, a reward of a man promoting the survival of her children would make sense; being rewarded with a rich man is a possible example. He can provide a guarantee of food, shelter, protection from war and pestilence, and more for her and her children. Perhaps money cannot buy happiness, but fitness can.

Unfortunately, when considering this case of gender bias, our altruistic woman is a unit of a superorganism; she is not a member of a modern society in which some men can be rich and others poor. The main point is that as a member of a superorganism, her survival needs and those of her children are met by society, not by a mate. Another point is that egalitarianism is a condition where all are equal. All men are equally rich, and society cannot reward her with a special man, as, other than altruists, they do not exist in egalitarian societies. Furthermore, any mechanism, including a reward of a special male(s), resulting in her children acquiring differential access to improved survival would violate the basic rule of egalitarianism. It would not happen.

This imbalance raises the question, what about the nonegalitarian aspect of altruistic men getting increased access? Egalitarianism seems to apply to the equal prospect for survival and is not concerned with differential reproductive advantages, which appear to be allowed. Given biology, this can only result in what we observe: increased fitness for altruistic men but not for altruistic women or their children.

Evolutionary consequences of apparent altruism. The foregoing provides insight into the evolution of a type of altruism independent of any requirement for commonality of genes as there is in kin selection. In this type, genetic relatedness plays no role. We can call it altruism, but a better name might be *apparent altruism* or even *compensated altruism*. Whatever its name, it is a type of altruism where fitness-decreasing acts performed for the group result in the group compensating with increased fitness. Importantly, the group we are discussing must be a superorganism, as only a biological entity can have fitness.

From an evolutionary perspective, altruism without some return reward for the altruist is a win–lose proposition, one that will go nowhere because there is no reciprocity. When the superorganism provides a compensation of increased sexual access in return for altruistic behavior benefiting the group, it changes the proposition to win–win. This is a zero-cost process because, one way or another, men will get access, else all comes to a stop. As stated earlier, all evolution is doing is making "free money" off what comes naturally.

For this process to work, the net cost in fitness to individuals probably must be zero or negative, as a consistent reduction in individual fitness leads to a dead end.

However, there is a downside risk or potential cost to the individual. Achieving the reward of increased fitness occurs only if the altruistic act is successful; evolution does not recognize intent, only results. Both *unokai* and *tokwaybagula* wannabes might do their absolute best and still not succeed, as the average gardener and so-so warrior get no such rewards. Evolution, like real life, gives no rewards for competing; it only gives rewards for winning. The evolutionary sharpening of pencils would not happen otherwise. For the Yanomamö warrior, failure could well be death—a not very fitness-enhancing outcome. The Trobriand Islander could simply have his Herculean gardening effort erased by a swarm of insects or hungry pigs breaking through the fence.

If evolution behaves in a rational manner, a greater downside risk should have a greater upside fitness potential. In the sexual sweepstakes game, a *unokai* should outcompete a *tokwaybagula*. The results of three experimental studies demonstrate the coupling between altruism and status. In reporting their results, Hardy and Van Vugt stated that "the most altruistic members gained the highest status in their group" and that "as the costs of altruism increase, the status rewards also increase."[6]

The big evolutionary payoff. The earlier discussed process for positive social control has a huge payoff. By rewarding the title *tokwaybagula,* the Trobriand Islanders became better gardeners, and with the title *unokai,* the Yanomamö became fiercer warriors. This is self-directed evolution; the population becomes what it wants. Its highest self-rated values are those becoming embedded in the population.

Self-direction is a double-edged sword. Consider that there must have been populations that did not value brave warriors or good gardeners. Where is the evidence of their existence? We will never know the answer. Populations had to decide wisely, as those who did not are no longer around to tell their tale.

Defeating cheaters, promoting altruists. There is a glaring problem associated with this system of apparent altruism. It depends on everyone doing her fair share. If someone works less than others do but still receives the usual share of labor's fruit, she gains fitness. Moreover, she does so at the expense of the fitness of others. This is the classic cheater problem associated with the evolution of altruistic behavior. Importantly, the superorganism cannot exist unless it solves this problem. It did, and in so doing, it established an interesting method to encourage altruists and concurrently repress cheaters.

If cheaters gain from the action of altruists and they themselves perform no fitness-decreasing altruistic acts, there is every evolutionary reason to be a cheater. In a population of cheater genes and altruist genes, cheaters always win. If we could not solve this problem, we would wind up, just like our cousins the chimpanzees, with societies of selfish cheaters, and altruism could never develop. Neither could the superorganism.

The definition of cheater includes "those who selfishly use common resources to maximize their individual fitness at the expense of a group."[7] In other words, cheaters decrease the group's fitness, making them the social antonym of altruists. It is in the group's evolutionary interest both to foster altruism and to repress cheaters.

In an earlier section, we pointed out that we do use group-approved sanctions to punish cheaters, but this is just one tool for fighting back pesky cheaters. There is another, less obvious and direct tool. It is rewarding altruistic behavior with an increase in fitness.

Cheaters do not walk the walk and gain no such rewards. Only altruists gain fitness from the group management of mating rules. When altruistic acts are so compensated, altruists become fitter than cheaters, resulting in their heritable genes and culture preferentially propagating. Those of the cheaters do not. This represents the leverage of *differential* reproductive fitness. In time, altruists, and not cheaters, are dominant in the population.

In theory, the only requirement to defeat cheaters is the differential treatment of altruists over cheaters. If altruists get nothing in return for their acts, while cheaters lose fitness, cheaters cannot take over. However, group rewards for altruists make that the win–win for both group and altruists. It is the starting point of the process. Inflicting punishment on cheaters increases process effectiveness but is not required. The key concept is differential fitness, with altruists winding up more fit than cheaters. That is all that is required.

Society, by managing mating rules, shifts the fitness advantage to altruists and away from cheaters. Society can and does praise altruists and punish cheaters, but the bigger and by far more effective hammer is the differential reproductive fitness of the altruist.

Nature of the Individuals of the Superorganism

Introduction. There are two perspectives of the superorganism. One is to look at the entirety of the organism, and that is what we have been doing. The second is to examine the individuals within, the ones who make up the group, and that is what we are going to do now. As with other areas of our story, there will be overlap between perspectives.

When we are a unit of a superorganism, we are not individuals, at least as individuals are commonly understood. The cocoon of the superorganism, of which we are an active component, subsumes our individual nature. In doing so, it constrains our behavior.

Being a member of a superorganism is not the same as being a member of the local booster club or an employee of a company. Being an integral part of a unit of natural selection is no small, trivial matter. After all, it is the target of natural selection and thus our interface with the evolutionary environment.

The following will be speculative for the reason that literature about group selection and the evolution of human psychology is difficult to find at best. For this, there are a couple of reasons. Firstly, as discussed, group selection is not an established proposition. Some scientific circles essentially forbid the concept, even though several highly respected scientists strongly advocate for it.

Secondly, the relationship of the individual to the superorganism is not a topic found in psychology literature. This is due to psychologists not normally considering Darwinian Theory as part of their paradigm. As a result, we have no guidelines for this part of our analysis and will have to draw inferences from the model we have developed.

The superorganism limits the individual's free will. Life for an individual as part of the superorganism was different from that of an individual in the modern world. The superorganism constrained the exercise of free will or agency, and we were less free then than we are now. This section is about that topic.

We will start with a bit of an academic exercise in sociology by discussing agency and structure. This will provide the framework for examining the behavior of individuals within the context of the superorganism. We will start with the following quotation:

> In the social sciences there is a standing debate over the primacy of structure or agency in shaping human behavior. Structure is the recurrent patterned arrangements which influence or limit the choices and opportunities available. Agency is the capacity of individuals to act independently and to make their own free choices. The structure versus agency debate may be understood as an issue of socialization against autonomy in determining whether an individual acts as a free agent or in a manner dictated by social structure.[8]

The superorganism comprises individuals who collectively decide what is acceptable and what is not. This means that the collective gives structure to agency, defining the limits to making choices. As members of the "deciding committee," we both prohibit and sanction agency. We collectively agree what individuals may and may not do.

These rules for agency are not ad hoc or spur of the moment. They accumulate over generations and weave the fabric of our culture's defining characteristics. It is from them that our customs and traditions originate.

Proximately, rules are established based on group concordance. Ultimately, many have as their basis evolution, with rules resulting in increased group fitness being prescribed and those in decreased group fitness proscribed. Settling on rules was probably the result of an evolutionary trial-and-error process, with surviving rules the result of group selection.

This is a feedback system. Selecting a good rule increases fitness and both the superorganism and the rule flourish. A bad rule results in the opposite. Survival is proof of good rules.

Some rules, perhaps most, have no effect on fitness. These add character but not substance to our culture. From an evolutionary perspective, they make us neither better nor worse, just interestingly different.

Groups making evolutionarily good choices survive, and so do their rules. A property of culture is to learn from and copy the successful rules of others. With imitation, good choices, like adaptive genes, spread.

Two differences between genes and culture are applicable to this discussion. One is that culture can propagate much more quickly than adaptive alleles. They do not require all that procreation stuff; all they need is a visit by a person exhibiting a different but better idea for living to start the process. The other is that, in this process, we can take another's culture, modify it to fit our particular condition, and make it our own. Culture is easily adapted; that is a second thing genes cannot do.

The superorganism reserves agency for its own use; this comes with being boss. Having agency, it can make changes. This is a requirement for effecting changes prompted by new environments.

Consequently, exercise of innovation or ambition is reserved for the superorganism. Any such behavior by the individual is not desired, as such actions can lead to internal inequality, an anathema in an egalitarian society. Actions or behavior resulting in change are the sole prerogative of the superorganism and are exercised only as part of the collective process.

Of course, the clever individual, as part of the superorganism, can influence its deliberations to an extent greater than those less clever can. For better or for worse, some individuals are effectively more equal than others are.

Reserving change for the superorganism has consequences. The individual making a change for the self normally risks only the self. There is not much risk to the group if an individual tries something new. When the collective does, the risk is much greater, and the entire group becomes at risk. Being reserved for just the group, change would be expected to occur infrequently and only in small increments. Groups having a propensity for large or frequent changes that deviate from the successful, historical norm would be expected to be less fit than conservative ones.

This part of our discussion suggests that the individual, when part of a superorganism, does not have agency or free will. This is not quite the case. The individual always has agency, but only if he is willing to suffer the consequences for its exercise. As part of a superorganism, if the individual did not like the rules of the group, he could leave. However, leaving could mean death, as the world exterior to the superorganism was hostile. Besides, we are ill designed to live alone. Individuals did not acquire individualism and unfettered agency until the Urban Revolution.

Custom and tradition as the basis for individual action. For a very long time, our evolutionary upbringing was confined to the cocoon of our superorganismic group. Over that time, we learned to conform to the group's desires, for if we did not, punishment awaited. We also learned to serve the needs of the group, and for this, we received sought-after praise. We evolved to seek praise and to avoid censure. In other words, we evolved to conform to the wishes of the group.

Culture informs us of the group's wishes. It teaches us what behaviors to avoid and the consequences for not avoiding them. It also teaches us how to behave to receive sought-after praise. Equally important, it teaches us how to respond to transgressions by others; culture informs what penalties we are expected to inflict on others for their transgressions.

A function of culture is the assurance of its heritance, and it does this by teaching every generation the same lessons. Culture may not be deterministic, as genes are, but its constancy across multiple generations makes it appear so.

We also indirectly learn of the wishes of the group through the history of tradition and custom, because implicit in this history is group concordance. The general rule is that you can do no wrong by doing what others are doing or have done. Observe and repeat. Thinking before acting is not only unneeded; it may not even be desired.

As noted, we lack literature specific for this topic. However, we can use the examples found in Malinowski's *Argonauts of the Western Pacific* to develop our arguments. This is the same source we used in chapter 6 for developing arguments supporting the concept of group selection. Because the Trobriand Islanders are a model superorganism, they are appropriate to use for examining the behavior of individuals who are part of a superorganism.

In the superorganism, custom and tradition dictate the actions of individuals. This is a point clearly made in the following quotations from *Argonauts of the Western Pacific*. While reading, keep in mind that these quotations are describing the nature of the superorganism:

> For, in every act of tribal life, there is, first, the routine prescribed by custom and tradition, then there is the manner in which it is carried out, and lastly there is the commentary to it, contained in the natives' mind. A man who submits to various customary obligations, who follows a traditional course of action, does it impelled by certain motives, to the accompaniment of certain feelings, guided by certain ideas. These ideas, feelings, and impulses are moulded and condi-

tioned by the culture in which we find them, and are therefore an ethnic peculiarity of the given society.[9]

The main social force governing all tribal life could be described as the inertia of custom, the love of uniformity of behaviour.[10]

In every community in the Trobriands, there is one man who wields the greatest authority, though often this does not amount to very much. He is, in many cases, nothing more than the *primus inter pares* in a group of village elders, who deliberate on all important matters together, and arrive at a decision by common consent. It must not be forgotten that there is hardly ever much room for doubt or deliberation, as natives communally, as well as individually, never act except on traditional and conventional lines. This village headman is, as a rule, therefore, not much more than a master of tribal ceremonies, and the main speaker within and without the tribe, whenever one is needed.[11]

Every man knows what is expected from him, in virtue of his position, and he does it, whether it means the obtaining of a privilege, the performance of a task, or the acquiescence in a status quo. He knows that it always has been thus, and thus it is all around him, and thus it always must remain. The chief's authority, his privileges, the customary give and take which exist between him and the community, all that is merely, so to speak, the mechanism through which the force of tradition acts. For there is no organized physical me by which those in authority could enforce their will in a case like this. Order is kept by direct force of everybody's adhesion to custom, rules and laws, by the same psychological influences which in our society prevent a man of the world doing something which is not "the right thing." The expression "might is right" would certainly not apply to Trobriand society. "Tradition is right, and what is right has might"— this rather is the rule governing the social forces in Boyowa, and I dare say in almost all native communities at this stage of culture.[12]

Custom and tradition are a flywheel of behavior. We have no lack of adages expressing this sentiment:

If it ain't broke, don't fix it.

Leave well enough alone.

Consensus doesn't have to change.

Don't rock the boat.

Never change a running system.

The success of the superorganism comes from unity of action, a function of group harmony. Disharmony leads to disunity, and groups having the greatest harmony have the strongest unity of action. They are the winners in the contest of group selection.

From an evolutionary perspective, selection should disfavor any behavior leading to discordance within the superorganism; maintaining harmony should be an evolutionary imperative. Not following custom and tradition can be disruptive and, consequently, disharmonious. Our nature should not be to change the routines of custom and tradition. If conservatism is resistance to change, then conservatism is probably our original state. It probably remained as such until the Urban Revolution and the initiation of psychologically modern man.

With the superorganism, selection pressures for traits fomenting deviation from custom and tradition should be low.[13] They are maladaptive for group fitness. These traits include innovation and ambition, and such traits in individuals are predicted to be low or nonexistent in pre–Urban Revolution societies. In the main, they are.

This does not mean that there is no change, only that change by an individual would be discouraged. If the majority so chooses, it can always deviate, as change to tradition and customs is the province of the collective. It is evolution by committee.

There is a second basis for the repression of innovation and ambition. Their exercise can lead to the individual becoming "better" or simply different from others, and this would result in a reduction in equality. Consequently, evolution should have selected against such behavior for as long as group selection persisted.

Conservatism is the preservation of social norms and includes those promoting adherence to the dictates of custom and tradition. It is a mainstay of the superorganism, as it is a positive force for its survival and continuance. Maintaining what works keeps it working. However, although it has a positive influence on the survival of the superorganism, it acts as a shackle on human progress. Let me explain.

As we have discussed, with the superorganism, change did happen, but only with approval of the group. Actions by individuals leading to changes in custom and tradition were discouraged. Resistance to change translates into change being slow.

Customs and traditions represent a history of success, and following them is a safe and sure posture. Changing them can be maladaptive for the superorganism and should be proscribed. This translates into constraint against the exercise of innovation or ambition by the group. It shackles culture to the past. Proscription of change maintains life, but only in a literally primitive sense.

A highly illustrative example of the shackle of conservatism associated with the superorganism is in Jared Diamond's book *The World until Yesterday: What We Can Learn from Traditional Societies*. In the quoted section, Diamond is describing life in the contemporary New Guinea highlands. This is an area that became exposed to the modern world only after its so-called discovery in the 1930s. Its people are probably still superorganismic, as they have not had the time to distance themselves from their recent Stone Age existence:

> As another example of traditional New Guinea society's de-emphasis of individual advantage . . . he was going to buy a sewing machine with which he would mend other people's torn clothes. He would charge them for repairs, thereby recoup and multiply his initial investment, and start to improve his lot in life. But Mafuk's relatives were outraged at what they considered his selfishness. Naturally, in that sedentary society the people whose clothes Mafuk would be mending would be people whom he already knew, most of them close or distant relatives. It violated New Guinea societal norms for Mafuk to advance himself by taking money from them. Instead, he was expected to mend their clothes for free, and in return they would support him in other ways throughout his life, such as contributing to his bride-price obligation when married.[14]

This is a shackle indeed! It is also an insight at the personal level into the nature of pre–Urban Revolution people, the poor of today.

As long as this shackle of conservatism remains, progress is restrained. Granted, this minimizes the risk of evolutionary demise and also its advancement. Taking current indigenous people such as the Yanomamö, the Trobriand Islanders, and the New Guinea highlanders as representative of nontransitioned people, this shackle explains some of their condition. In today's world, populations that remain shackled are a constituent of the poor.

In making the transition of the Urban Revolution, natural selection changes from group to individual, unshackling the individual from tradition and custom. The individual becomes a free agent and can exercise ambition and innovation. Both are major factors in human progress.

In a subsequent chapter, we will discuss the nature of the change in individual identity with the Urban Revolution. We will see that the major change is a shift from group to individual selection. It is a shift from constrained to unfettered individual agency. As long as group selection exists for a population, this constraint operates as a shackle on the collective of the superorganism. It keeps the primitive village primitive and has huge consequences for the wealth and poverty of nations.

The conservatism we have been discussing is not the usual form of "conservative" versus "liberal." It is one required by the superorganism and is innate to the superorganism and group selection. It does not apply to the modern world, where individual selection reigns, defining a demarcation between our earlier superorganismic being and our current modern one. It is the dividing line between the poor and the rich, a line between rigid conformity of the group and potentially unfettered change by the individual. It is one lending inertia to the old ways, anchoring the present, untransitioned people to the past.

CHAPTER 9

Consequence of the Superorganism

We have discussed the nature of the superorganism; now it is time to discuss its consequences. There are many, but we are going to focus on just three important to our analysis. The first is about the origin of ethnicity, and this discussion is part of the general one examining why people are different. The last two are specifically concerned with the reasons for the superorganism not performing well in the modern economy. With these, we will learn some of the reasons for superorganismic societies being poor.

Making Ethnicity

Introduction. We live in a world filled with a kaleidoscope of different ethnicities. Some are relatively small groups, like the Houma Indians of Louisiana, having a population of approximately seventeen thousand.[1] Others have huge representations, such as the Igbos of Nigeria, with a population of 32 million.[2] Throughout the world, there must be many thousands of such groups, both big and small, and each is unique. In this section, we are going to examine where they all came from, especially the source of their uniqueness.

Our story is about people becoming different, and that is what "making ethnicity" is. However, as a singular topic, it is very large, and covering it in depth would detract from our mission. In this section, we only want to provide the briefest of overviews, as this is all our analysis needs.

Defining culture. Our narrative is about people becoming different. We are going to see that this is a result of each superorganism creating its own unique culture. This makes culture obviously important, and we need to have a definition of such a property. There are several. Bates and Plog, in their book *Cultural Anthropology,* offer the following:

> Culture: The system of shared beliefs, values, customs, behaviors, and artifacts that the members of society use to cope with their world and with one another, and that are transmitted from generation to generation through learning.[3]

Another comes from *The Origin and Evolution of Culture* by Boyd and Richerson:

> Culture is information capable of affecting individuals' behavior that they acquire from other members of their species by teaching, imitation, and other forms of social transmission.[4]

Both of the preceding definitions are pertinent to our discussion, but for our purposes, we can best define culture as the effective genome of the group. This places it into the category of the phenotype, a selectable property in natural selection.

Acquiring culture. Earlier, when we discussed social control and desired behavior, we talked about an evolutionary system.

Group preferences leading to group-rewarded status in turn leading to reproductive fitness is a powerful mechanism. In a like manner, there is also decreased reproductive fitness associated with behavior disapproved by the group. If there is a single message in our story as to why people are different, this is it. Importantly, the driver is culture.

We now have a complete system, one that is majority rule and bottom up. The group decides what behavior it likes and dislikes. It has tools to prevent one and promote the other. This is an evolutionarily powerful machine. Our ability for social control is what gives us culture and morals, traits not found in other animals.

The consequence of the aforementioned process is the development of a culture unique to the superorganism.

This same group-centric process adds to our customs and traditions. We habitually do things that we enjoy or are effective—no point repeating things that are not. These become the flywheel of our behavior, and it is our conservatism that keeps them running. The superorganism is a machine with inertia.

Our superorganism is not a point source of ethnicity. We did not achieve it all by ourselves, nor did we keep it to ourselves. We had help, and we shared. Neighbors and casual acquaintances whose traits we found useful, we copied, and they likewise copied ours. The people and societies we copied were not very distant; we had no means to travel very far. This kept any culture local, even if it became distributed among several groups. Another consequence of this constraint is that culture evolves in a local, specific environment.

Another source of change was the spoils of war. At times the defeated who survived became slaves to the victors. Independently of being slave or free, the winners often incorporated losers into their body victorious. This makes the consequence of war a mechanism for both cultural and genetic change.

This entire process created group-specific culture, the root source of ethnicity when it is defined by a common psychology. Each superorganism had its environment. It also had a developmental history on which to build, layer by differing layer, changes and additions to the culture that existed. That combination of environment and history for any group was unique and led to each group having its unique identity. This is an identity defined by a common physiology and psychology. This is ethnicity.

Over the vast span of time, superorganisms come, go, combine, and fission. It is with the earlier noted process that, over this time, they changed and became different from one another. Our ancestral groups were not static; they changed with their adapting cultures and environments. They became us, different in innumerable ways, both physically and mentally.

Culture is selectable. Culture is a property of the superorganism, and in group selection, it is a selectable variable. Naturally, it selects the groups with the "best" culture. "Best" is the determinant of survival and is dependent on the environment.

Culture, including morality, is a human construct. By their choices of cultural preferences, groups were the architects of their evolutionary fates. Wittingly or not, our ancestors chose their own evolutionary futures. Natural selection chose superorganisms making "wise" choices about how they wanted to live. Those making evolutionarily wrongheaded choices were either eliminated or adsorbed by those who did not.

Groups in group selection, just like individuals in individual selection, undergo differential survival. When cultural variability between groups is the basis for such selection, it is termed *cultural group selection*. Boyd and Richerson state, "Cultural group selection is analogous to genetic group selection but acts on cultural rather than genetic differences between groups."[5] Additionally, a property of group culture and group selection is that "cultural evolution also creates new selective environments that build cultural imperatives into our genes."[6]

Remember, natural selection operates on the phenotype. Culture has no direct role in the phenotype's parent, the genotype. What it does is modify the environment, thus making a phenotype different from the genotype. This effectively makes culture a variant for selection.

In other words, culture creates and molds the environment for selection, making culture not only a property of the group but also a product. We designed our future selves and evolution made what we designed. Man makes culture; evolution selects; survivors make new culture. The process is a change–test–modify one, which builds an entity layer by layer.

The myriad cultures we find today, such as the Igbo of Nigeria, the Pashtuns of Afghanistan and Pakistan, and the Ainu of Japan, are those that provided their group more fitness than their evolutionary competitors did. They are all success stories because they survived. Their competitors did not, a process separating the good from the bad. Currently there are no bad cultures; by definition, they no longer exist. In the ultimate analysis, value judgments of this sort are nature's province, not man's.

Economy and the Superorganism

Introduction. When we were chimpanzee-like, living in Boehm's ancient forest, did we have an economy? If chimpanzees do not have an economy, then we, too, at that distant time, had none. Sometime between then and now, we got "economy." How and why we did is an important part of our analysis.

We are going to examine the superorganism in the context of economy, but we are not going to get to where we actually have "economy." We just want to build the foundation so that, when we later discuss our "getting economy" as a consequence of the Urban Revolution, we have something to stand on.

We open the conversation on this topic with two quotations providing insight into the economic nature of pre-state societies:

> The outstanding discovery of recent historical and anthropological research is that man's economy, as a rule, is submerged in his social relationships. He does not act so as to safeguard his individual interest in the possession of material goods; he acts so as to safeguard his social standing, his social claims, his social assets. He values material goods only in so far as they serve this end. Neither the process of production nor that of distribution is linked to specific economic interests attached to the possession of goods; but every single step in that process is geared to a number of social interests which eventually ensure that the required step be taken. These interests will be very different in a small hunting or fishing community from those in a vast despotic society, but in either case the economic system will be run on noneconomic motives.[7]

> In such a community the idea of profit is barred; higgling and haggling is decried; giving freely is acclaimed as a virtue; the supposed propensity to barter, truck, and exchange does not appear. The economic system is, in effect, a mere function of social organization.[8]

Clearly economics for the superorganism is different from economics for modern man.

The rational actor. There is in the discipline of economics the concept of the individual as rational actor. According to this theory, individuals behave in their own self-interest, and, being part of economic theory, self-interest is often decided from an economic perspective by economists.

We are rational actors, but the reason for this is within biology. For organisms having behavior, rationality is mainly about getting genes into the future. Sometimes that is coincident with an economic perspective, sometimes not. Rational behavior should normally be "fit behavior," that is, behavior promoting survival and reproduction.

In the preceding quotation, individuals act to preserve their social standing, social claims, and social assets. This is behavior an economist might not consider as rational, as there are no economic ends for the labor. However, as we will explain in this section and throughout our story, it is very rational. It is ultimately about getting genes into the future.

We have already discussed something about economics and the superorganism, just not explicitly. We did it implicitly in our discussion of labor for the superorganism and altruism. Recall that when we lived in superorganismic groups, much of our labor was altruistic; it was for the good of the group. This altruistic labor was only apparent. For fitness-decreasing labor, we received fitness-increasing compensation. What appears to be irrational altruistic behavior is actually rational, as there was substantial reward for the labor.

When we labored for the group, our remuneration was status. As we have discussed, status is a fitness indicator, making status itself a factor in fitness. Individuals with high status are fitter than those with low status. Therefore we labored for our reproductive fitness—a biological, not economic, reward.

Obviously, rationality can have a basis other than an economic one. In the broad scope of things, rewards of fitness have greater value than rewards of wealth. Getting genes down the road is a greater reward in an absolute sense than a bigger house or fancier car is.

Needs. We first need to understand "economy," and for this we can start with the definition of the Economic Problem:

> The economic problem—sometimes called the basic, central, or fundamental economic problem—is one of the fundamental economic theoretical principles in the operation of any economy. It asserts that there is scarcity; that is, that the finite resources available are

insufficient to satisfy all human wants and needs. The question then becomes how to determine what is to be produced, and how the factors of production (such as capital and labor) are to be allocated. Economics revolves around methods and possibilities of solving this fundamental economic problem.[9]

At its basest level, "economy" is about scarcity and not being able to meet wants and needs. To develop our understanding, we are going to divide this discussion into needs and wants.

Needs are things people require for survival. They include food, shelter, and clothing. Needs are finite; once satisfied, more are not required. *Wants* are things in excess of needs. You do not have to satisfy a particular want to survive and reproduce. You might need a house, but a grander one is simply a want. Protein is a need; a steak is a want. We could go on making lists of various types and classes of wants, but the main issue is that they are infinite. This is where economics comes in. It formalizes the problem of infinite wants and the scarcity of means to satisfy them.

The next step in developing our argument is to resolve needs and wants into their evolutionary components. Needs are things we must have to survive and reproduce; wants are similar, but they fall into the category of only "nice to have." We need food but can only want steak.

Evolution is pragmatic; it only solves real problems. The superorganism, like all organisms, is a consequence of evolution. Its evolutionary design was concerned only with satisfying needs, as satisfying wants would not be adaptive. For the very long evolutionary time of our superorganismic existence, we had no wants. We had to wait for the Urban Revolution to acquire a taste for them.

Work to meet needs. We will start this part of our analysis of needs, work, and the superorganism with *Argonauts of the Western Pacific*:

> Another notion which must be exploded, once and forever, is that of the Primitive Economic Man of some current economic text books. This fanciful, dummy creature, who has been very tenacious of existence in popular and semi-popular economic literature, and whose shadow haunts even the minds of competent anthropologists, blight-

ing their outlook with a preconceived idea, is an imaginary, primitive man, or savage, prompted in all his actions by a rationalistic conception of self-interest, and achieving his aims directly and with the minimum of effort. Even one well established instance should show how preposterous is this assumption that man, and especially man on a low level of culture, should be actuated by pure economic motives of enlightened self-interest. The primitive Trobriander furnishes us with such an instance, contradicting this fallacious theory. He works prompted by motives of a highly complex, social and traditional nature, and towards aims which are certainly not directed towards the satisfaction of present wants, or to the direct achievement of utilitarian purposes. Thus, in the first place, as we have seen, work is not carried out on the principle of the least effort. On the contrary, much time and energy is spent on wholly unnecessary effort, that is, from a utilitarian point of view. Again, work and effort, instead of being merely a means to an end, are, in a way an end in themselves. Several of my friends, renowned as *tokwaybagula,* would boast to me how long they worked, how much ground they tilled, and would compare their efforts with those of less efficient men. When the labour, some of which is done communally, is being actually carried out, a good deal of competition goes on. Men vie with one another in their speed, in their thoroughness, and in the weights they can lift, when bringing big poles to the garden, or in carrying away the harvested yams.[10]

Why does the Trobriand Islander work? At the first cut, economists would have them working to satisfy basic needs, such as obtaining food and shelter. However, from the preceding quotation, we learn that they work for status; that is what competing to be a *tokwaybagula* is all about. So, how do we get from achieving *tokwaybagula* status to meeting the needs of the islanders? The answer is in how the "economy" of the superorganism functions.

Garden work in the Trobriand Islands produces yams, the islanders' primary staple. In their system, gardeners do not reap what they sow, and the harvest is distributed to others, usually in-laws. However, the net consequence of their social rules for food distribution ensures sustenance for all. As long as there are yams, no one starves.

If the social system meets the food needs of all, is there any reason to work hard or even work at all? Is there any reason not to be a slacker? The answer is yes, because of differential reproductive fitness. As we have discussed, when altruists receive fitness rewards, cheaters lose out. This is the consequence of good gardeners in the Trobriand Islands receiving the reward of *tokwaybagula*.

We cannot tell from Malinowski's opus how receiving the title of *tokwaybagula* increases fitness, but there is an educated speculation. We know that the gardeners are male and that gardening is a competitive exercise. We also know that the root of male competition is often access to women. The speculative proposition is that the group reward of *tokwaybagula* confers on the recipient a leg up in the sexual sweepstakes contest, and they wind up with more progeny. Good gardeners get girls.

Wants. In the context of the economic problem, wants are things in excess of needs and are unlimited. For our purposes, we are going to assume that effort for wants only occurs once needs are met.

Humans, like all living things, have always had needs. Can we say the same about wants? If we return to our starting reference, the chimpanzees, we see that if they do have wants, it is not very obvious, and if so, very little. This means that at some time in the past, neither did we. With modern humans, our wants are just the opposite of the chimpanzees: obvious and more than we probably should. Somewhere on our journey to modern man, we got wants.

There is an interesting question: how do we know if a society has wants? We know they have needs; it goes with their biological territory. Because biology does not require wants, there may not be any direct evidence for their existence. We will have to use a little deduction.

Satisfying needs and wants requires human effort; you have to work for them. If a society has a difficult time satisfying needs, we probably would have no knowledge if they had wants. Assuming their presence, it is only after needs are met and excess production capacity still exists that we will observe wants.

We are going to deduce that if there is substantial, unused labor available for work after the completion of necessary chores for needs, and it is not used for meaningful production, there are no wants. For instance, a society with only a twenty-hour workweek used primarily for meeting needs probably has no wants.

We have used Chagnon's *Noble Savages* as our reference for the Yanomamö. We pointed out in our discussion of their lifestyle that after working in the morning to meet the needs of the tribe, women passed the time gossiping. Men, after completing their chores for the same purpose, used their free time to take hallucinogenic drugs. Nowhere do we see the Yanomamö working for wants. The only work we see is done to meet needs. Likewise, in *Argonauts of the Western Pacific*, there is no evidence of the Trobriand Islanders working for wants.

Neither the Yanomamö nor the Trobriand Islanders show evidence of food scarcity. They do not appear to have any problems in meeting their sustenance requirements. They lack neither time nor resources and could easily produce more if desired.

If they were part of a modern society, both of these groups would spend available time, after the production of items for needs, working to satisfy wants. Possible examples might be making a bigger hammock or extra chairs. They would work to "improve their condition," and if they did not, their peers might consider them indolent. Neither the Yanomamö nor the Trobriand Islanders work for wants, assumedly because they have none.

Granted, some societies, like forager societies, had to maintain ease of mobility and could not acquire stuff. They could not afford to lug around the consequences of material wants. However, neither the Yanomamö nor the Trobriand Islanders were migrant foragers. The Yanomamö only moved every two or so years, and the Trobriand Islanders were stuck on their respective islands. Both groups appear to be sufficiently sedentary, where wants hindering mobility would not be a burden. Nevertheless, nowhere with these two groups do we see wants.

The following quotation provides further evidence. It is from a monograph on the economy of the Kapauku Papuans, a people of the Indonesian province of Papua, a province comprising most of the western half of the island of New Guinea:

> Since the Kapauku have a conception of balance in life, only every other day is supposed to be a working day. Such a day is followed by a day of rest in order to "regain the lost power and health." This monotonous fluctuation of leisure and work is made more appeal-

ing to the Kapauku by inserting into their schedule periods of more prolonged holidays (spent in dancing, visiting, fishing, or hunting; in the Kamu Valley the last two activities are considered recreation). Consequently, we usually find only some of the people departing for their gardens in the morning; the others are taking their "day off." However, many individuals do not rigidly conform to this ideal. The more conscientious cultivators often work intensively for several days in order to complete clearing a plot, making a fence, or digging a ditch. After such a task is accomplished, they relax for a period of several days, thus compensating for their "missed" days of rest.[11]

Another example of unused labor is with the Kuikuru, a people occupying a single village of approximately two hundred persons in the upper Xingu River area of central Brazil:

There is no doubt at all that the Kuikuru could produce a surplus of food over the full productive cycle. At the present time a man spends only about 3-1/2 hours a day on subsistence—2 hours on horticulture, and 1-1/2 hours on fishing. Of the remaining 10 or 12 waking hours of the day, the Kuikuru men spend a great deal of it dancing, wrestling, in some form of informal recreation, and in loafing. A good deal more of this time could easily be devoted to gardening. Even an extra half hour a day spent on agriculture would enable a man to produce a substantial surplus of manioc. However, as conditions stand now there is no reason for the Kuikuru to produce such a surplus, nor is there any indication that they will.[12]

Sahlins, in his book *Stone Age Economics,* provides several examples of unused labor in primitive societies. In those societies, labor is for sustenance and householding, never for satisfying wants. Sahlins states that primitive societies were affluent, where the metric for affluence was not the amount of material goods accumulated but rather the amount of leisure time available for nonproductive activities.[13]

Superorganismic societies used leisure for loafing, gossiping, social entertainment, and sleeping, not for improving one's status in life. Egalitarianism is all about *not* improving individual status; else, it would not be egalitarianism. Even attempting to improve one's status was probably a social no-no in these societies.

The superorganism has needs but lacks wants. Given that the economic problem is one of managing needs and wants, is there "economy" if there are no wants? Perhaps, but that is properly a question for economists to answer. For us, it is not an important question; its answer will not change our direction or understanding.

However, viewed from a different perspective, it is *the* critical question of our narrative. Without wants, can a society economically prosper in the modern world? When referenced to any modern, advanced nation, the answer is no. You really have to want stuff to be prosperous by that standard.

A conclusion of our story will be that there are populations who are or have remained until recent times without wants. We will later discuss that a consequence of not having wants is a low standard of living, but only when compared to modern standards. In the modern world, you have to want material goods to have the good life! The important point for our story is that populations lacking an economic history of wants are a major constituent of the poor of today's world.

Later, we will discuss the evolution of wants and its consequence, the modern economy. For now, we are stuck with the superorganism, human needs, and life without stuff. At this point in our narrative, we remain the *Homo bellicus* of an earlier chapter; *Homo economicus* awaits.

Work, Indolence, and the Superorganism

Introduction. We have stated throughout this story that many of today's poor are those either not making the transition of the Urban Revolution or only just recently doing so. This part of our discussion will be about those nontransitioned populations in modern or near-modern times. For this, we are going to leave the theoretical ivory tower and get out among the folks. It is time to see what these people are like and to couple our model of the superorganism to people we can feel, see, and touch. It is time to understand why they are poor.

Insightful work. If we were to revisit the first chapter, we would see that, according to the historian Landes, living to work is the progenitor of prosperity. The other side of that coin, working to live, might make people happy, but it will never make them rich. The difference expressed in these two attitudes about work offers insight into our original question of why some nations are rich while others remain poor.

In the intervening pages since chapter 1, we have developed a picture of pre-state man as an entity within a superorganism. To arrive at an answer to our original question as to why there are rich and poor, we need to understand the superorganism from the perspective of work and the work ethic.

The following portrait of the superorganism's economic structure is from a paper by Malinowski that discusses the economics of the Trobriand Islanders. It provides a starting point:

> In savage societies national economy certainly does not exist, if we mean by the term a system of free competitive exchange of goods and services, with the interplay of supply and demand determining value and regulating all economic life. But there is a long step between this and Buecher's assumption that the only alternative is a pre-economic stage, where an individual person or a single household satisfy [sic] their primary wants as best they can, without any more elaborate mechanism than division of labour according to sex, and an occasional spasmodic bit of barter. Instead, we find a state of affairs where production, exchange and consumption are socially organised and regulated by custom, and where a special system of traditional economic values governs their activities and spurs them on to efforts.[14]

This describes an environment where work's reward is "a special system of traditional economic values" and not any remunerative or other usual economic reward. It is for satisfying values set by custom and tradition. It is not work for its own sake or work simply to meet some mundane ends. It is what we have said earlier, for social not economic reasons, and this is what spurs the individuals of the superorganism to labor. Clearly work in Malinowski's sense is something other than that which Landes described back in chapter 1.

Framing our view. The foregoing opening to this part of our discussion is correct, but a bit misleading. It is improperly framed in a Eurocentric perspective, one based on individual selection. Properly, the frame should be group selection. It is only in that sense that labor in primitive societies, as we have described here and elsewhere, makes sense.

The work Malinowski discussed is internal to the group sur-vival machine, the unit of natural selection. We have to view the entirety of the effort of the individuals of the collective to under-stand its effect and role. Internally, individuals work for status, but effectively that collective effort sustains the group. The totality of the work system meets the needs of the entirety at both the indi-vidual and group levels.

This type of egalitarian system, with an internalized economy, survived and has done so for millennia. It will never be rich. It also will never be poor. It has no need for "poor" or "rich"; they are *non sequiturs* in such a system. They simply do not apply to the superorganism's egalitarian and self-contained internal frame of reference. It is only when we shift our reference frame to the usual economic one of the individual, self-interest, and hierarchy that we have *rich* and *poor* as useful, relative terms.

Our analysis requires the comparative basis of rich and poor. That is what we are about. We need the perspective of the Euro-centric frame, even though, at one level, it is misleading. When we take this perspective, we see that the superorganism is poor. That is its condition upon entering the portal of the Urban Revolution.

The colonists and the generic "lazy" natives. A major reason for European colonization of what at that time they considered the uncivilized world was economic. New lands offered economic op-portunities, as there were new resources for exploitation. Impor-tantly, there were primitive men for labor, and because they were primitive, their unskilled labor was expected to be cheap. With un-tapped resources and cheap native labor, there was lots of money to be made.

That might have been the plan, but as we will see, unexpected reality can get in the way, especially when it is incomprehensible. The colonists had not a clue.

In the following quotation, the economist Polanyi discusses the problem of getting wage labor in colonial Africa:

> The natives are to be forced to make a living by selling their labor. To this end their traditional institutions must be destroyed, and pre-vented from re-forming, since, as a rule, the individual in primitive society is not threatened by starvation unless the community as a

whole is in a like predicament. Under the *kraal-land* system of the Kaffirs, for instance, "destitution is impossible: whosoever needs assistance receives it unquestioningly." No Kwakiutl "ever ran the least risk of going hungry." "There is no starvation in societies living on the subsistence margin." The principle of freedom from want was equally acknowledged in the Indian village community and, we might add, under almost every and any type of social organization up to about the beginning of sixteenth century Europe, when the modern ideas on the poor put forth by the humanist Vives were argued before the Sorbonne. It is the absence of the threat of individual starvation, which makes primitive society, in a sense, more human than market economy, and at the same time less economic. Ironically, the white man's initial contribution to the black man's world mainly consisted in introducing him to the uses of the scourge of hunger. Thus, the colonists may decide to cut the breadfruit trees down in order to create an artificial food scarcity or may impose a hut tax on the native to force him to barter away his labor. In either case the effect is similar to that of Tudor enclosures with their wake of vagrant hordes.[15]

The preceding quotation is consistent with the understanding of the superorganism we have developed. There are no surprises, and it nicely describes the misfit nature of the superorganism as it relates to a modern economy. The natives felt no threat of individual starvation because their social organization met their subsistence needs. They had freedom from want because they had none. For them, wage labor, besides being unknown, served no purpose.

With the natives not having any concept of wage labor, there was no obvious inducement to get them to labor. That had to be created. The colonists, in order to get native labor to meet their commercial needs, first had to destroy their means of obtaining subsistence. Only once so deprived would they perform wage labor. Colonization not only subjugated, it cruelly deprived the natives of the only means of livelihood they knew. It was either labor for the new masters or starve.

This may not meet the technical description of enslavement, but it comes very close. There is more, and what Polanyi does not mention is that once coerced to wage labor, they would only work to the extent required to meet subsistence needs and no further. Without wants, an understanding of which the Europeans had not a clue, the natives stopped work once their bellies were full. Because the colonizing Europeans lacked any understanding of this attitudinal phenomenon, they considered the natives lazy.

The colonists could well have muttered a paraphrase of Professor Henry Higgins's lament from *My Fair Lady*: "Why can't the natives be more like us?" Ignorance was not bliss for any side.

Indolent Indonesians. Ronald Seavoy lived and worked in Indonesia in the second half of the twentieth century. Observations in his book *Famine in Peasant Societies* bring the superorganism of our story into modern times. The Indonesian peasants he describes do not appear to have made the transition of the Urban Revolution, and we will use them as a prototypical, nontransitioned society living in modern times.

Seavoy has obvious biases, and we must carefully parse his writings to avoid biased conclusions. His main bias is that natives are indolent, a somewhat common Eurocentric perspective of the colonial era.

The *Oxford English Dictionary* defines *indolent* as "wanting to avoid activity or exertion; lazy," and on this Seavoy is wrong. However, Seavoy expands the definition to include "proactively minimizing work," and on this Seavoy is correct.

We will use his observations and our model of the superorganism to explain what the natives actually are. In doing so, we will find the natives not to be lazy, but at the same time, they proactively minimize work. However, this is not for Seavoy's explanation of indolence. Our model will explain this disconnect.

The list of ethnic groups comprising these rural Indonesian villages includes Javanese, Sudanese, Malay, Batak, Balinese, Madurese, and Buginese, and the religions include Islam, Christianity, Hinduism, Javanese (Abangan), and animism. Each village speaks its own dialect. They mainly subsist on rice, maize, cassava, sago, and palm sugar.

Seavoy states that the basic social unit of all peasant societies in Indonesia is the village and that the village employs subsistence agriculture. This is a mode of production where the peasants produce for their own consumption and very little, if any, production is for trade or commerce. This means that the village is self-contained and has little or no need for economic activity with others.

Commercial farming differs from subsistence agriculture in that their respective objectives are different. The objective of commercial farming is to optimize the return on investment. One way of measuring this is with calories per hectare, where the land under cultivation is the investment and calories have a monetary value. The cost of labor is a factor.

Subsistence farming has a different set of objectives. Its primary objective is to produce enough calories to meet per person needs. There is a secondary one: the subsistence farmer wants to minimize the amount of labor required to produce the calories. He does not want to work any more than required. The metric is calories per unit of labor. Unlike commercial farmers, and for reasons we will later explain, profit is not a motive. Once the farmers meet their per person requirements, labor ends.

For Indonesian and other similar slash-and-burn farming communities living in remote regions and with low population densities, land is often not a factor. Once their growing activities have depleted the soil's nutrients, they pick up stakes and move on to a new plot. Their only cost is land preparation, where clearing forests can be a major effort.

In Indonesia, subsistence-based villages are relatively isolated and, of necessity, self-sufficient. Households in these villages produce 95 percent or more of goods and food consumed. In the main, commercial agricultural production is not known, but there are cases when the peasants are forced to produce commercial products, such as copra or rubber, for landowners. When this happens, they often produce only enough to meet subsistence needs:

> Higher wages . . . did not result in an increased supply of labor but, on the contrary, in a decreasing supply. . . . The needs of a village family could, at that time, be met with only a very small amount of cash. Financial wealth did not itself confer prestige in the village community so that no one was particularly keen on having cash on hand. . . . The villagers quit working at the plantations when they had earned enough cash for their tax, salt, kerosene, and clothing.[16]

The foregoing is consistent with working just to meet needs. If you want the natives to work more, you have to get them to want, to become desirous of the material rewards of work.

In Indonesia, several types of agriculture are available for subsistence farming, ranging from slash-and-burn cultivation to labor-intense, terraced wet rice production. We can use two criteria for evaluation. One is the amount of labor required per calorie produced, the other the amount of food or number of calories produced per unit area.

Slash and burn requires the least amount of labor and produces the least number of calories per unit area. Terraced wet rice production is at the other end of the spectrum, requiring high labor input but yielding a high number of calories per unit area.

An important variable for consideration is the amount of land available for cultivation on a per person basis. Seavoy reports that when there is adequate available land, the villages choose slash-and-burn techniques because this meets their needs with a minimal level of effort. They optimize on calories per unit of labor.

When population densities are high, villages must resort to higher-labor methods that produce a greater number of calories per unit area to meet sustenance requirements. The highest population densities use terraced wet rice methods. At some point, the local population density exceeds the capacity of the land, and once that condition is met, some move from the village to the city to engage in wage labor for meeting their needs.

Seavoy argues that they choose slash-and-burn methods over terraced wet rice cultivation because the natives want to optimize indolence. In this assessment, he could be correct, but if he were an engineer or population biologist, he never would have made such a statement in the first place. Rather, he would have made the argument that they chose the most energy-efficient means of meeting sustenance needs. That is what optimizing on calories produced per calorie of labor is all about.

Choosing the method yielding the maximum number of calories for calories expended is what successful organisms do. Even in biological systems, energy efficiency has a positive payoff, one reflected in the outcome of natural selection. Do the Indonesian village peasants choose a method for maximizing energy efficiency or satisfying indolence? The results satisfy both motives, leaving the answer to depend on the education and biases of the questioner.

Feeding social needs. In Indonesian peasant villages, just as in similar villages throughout the world, people work primarily to satisfy needs. As in the Trobriand Islands, status for such work can be a motive. However, the main point is that there is no work for wants. Once work for needs is completed, village residents pursue other activities, such as socializing, sleeping, or fishing for recreation.

Remember, from earlier in our discussion, that "superorganism" is an upscale name for "village." With the peasant village, we are dealing with an organism, not simply a collection of huts in a forest clearing. In these villages, is social activity a requisite for maintaining the social health of the superorganism? I suspect that it is. It could be that there exists with superorganisms a need for some level of social activity.

If true, some of what appears to be indolence is actually time spent in social activities needed for group maintenance. This could be called *social grooming* and could be in the form of gossiping, fishing with friends, or even taking drugs. As long as it is social, it is of benefit to the group. Monastic activities, such as reading or sleeping, are not. Indolence, provided it includes activities that are social in nature, might have value.

Socially maladroit wealth. In modern societies, labor in excess of needs is often for acquiring wealth. This is not so in egalitarian ones. In such societies, becoming wealthy is socially maladroit, and such societies discourage individual wealth creation. This point is validated in the following quotation about the peasants of Indonesia:

> Peasants do not place a high social value on commerce because it creates men who are relatively wealthy and who are not under the control of customary laws administered by village councils. The social goal of the peasantry is not to improve its material welfare, but to preserve a social order that allows it to subsist with minimal labor expenditures.[17]

The peasants of Indonesia and elsewhere have survived for millennia with their work ethic along with their customs and traditions. Resisting change and thereby preserving social order makes evolutionary sense. Change gets to new, untested, and therefore possibly dangerous waters. If maintaining the social order of the superorganism had not been a priority of our ancestors, we, you and I, would probably have a different psychology. We would be different from what we are.

Work's reason. For everything exists a reason, including work. Reasons usually have rationality and can differ between cultures. For the Yanomamö, work was for meeting subsistence needs; there does not appear to be any other reason for it. After fulfilling the reason for work, individuals were free to do whatever they liked. Men took drugs and women gossiped. For the Trobriand Islanders, reasons for work are a little more complex. The proximate is to earn the title of *tokwaybagula* while the ultimate is to provide subsistence for the tribe.

The European colonists came from a different world, one of unlimited wants. They wanted a change of clothes, a servant or two, or perhaps a big enough dowry to snag an important man for an unmarried daughter. We could make this list substantially longer, but it makes the point. They had unlimited requirements for work's reward: wealth. The colonist's cultural norm was to work diligently to satisfy these unlimited wants, and as part of the process, they got enough to meet needs. Their problem was prioritizing. They had to choose what work to perform. For whatever labor they chose, leisure for them, and their culture, was for the indolent and lazy.

The colonized lived in a different world, one of no wants, only needs. Even though their motive could be status, satisfying needs was their only reason for work. Needs are unlike unlimited wants; they have natural limits. You can only eat so much and live in one house at a time. Once individuals of the superorganism fulfilled needs, they had no other reason for work. They worked to live, and once their needs were met, further work was not required.

Is a man lazy if he does no work, given that all reasons for work have been fulfilled? Who or what establishes requirements for work? Who decides on lazy? Who determines poor? Who gets to judge others?

They Are Different

The largest island in the world is Greenland. The second, lying just to the north of Australia and at the end of the Indonesian archipelago, is New Guinea. Up until the early 1930s, the world only had knowledge of those living in the relatively accessible coastal areas. Then, in the inaccessible central highlands, the modern world discovered hundreds of tribes still living in Stone Age conditions. From that relatively recent time of first contact until now, various NGOs, government organizations, religious groups, and others, even including tourists, have been bringing the inhabitants of this remote piece of land into modern times.

Jared Diamond, author of *Guns, Germs, and Steel,* has made several journeys into these newly exposed lands and has been eyewitness to the nature of these prehistoric, traditional societies. From these travels, he wrote the book *The World until Yesterday: What We Can Learn from Traditional Societies.*

A pertinent impression of who we were as individuals of a superorganism comes not directly from Diamond's book but rather from a *New York Times* book review. It contains a sense of people who are different from the modern us, of people alive in the modern era but still living in a most distant time, one existing well before any Urban Revolution:

> The people Diamond describes seem immersed in the collective. We generally don't see them exercising much individual agency. We generally don't see them trying to improve their own lives, alter their destinies or become a more admirable people. It's possible they do not conceive of life in this individualistic way. It's possible they don't see life as a journey, as we tend to, but as a cycle. It's possible they don't conceive of history as having direction, as we tend to, but just as an endless return.
>
> Many books have been written comparing the hyper-individualism of today's society with the more communal patterns that have been left behind. It is hard to learn from this one because the traditional people, at least as described here, feel so different from us.

> They don't seem like us, the day before yesterday. They seem like people separated from us by large chasms—by all the events of our written history, all the ideas of our thinkers, all the teachings of our religions.

Much of the review includes topics we have discussed during our development of the superorganism. It gives them a context apart from the abstract ivory tower. Importantly, it paints a verbal picture of our ancestors before the formation of cities, the advent of civilization, and the rise of the state:

They are immersed in the collective.

They do not exercise much individual agency.

They do not try to improve their own lives.

They do not try to alter their destinies.

They do not try to become a more admirable people.

And, the concluding, salient point:

They do not seem like us.

We, you and I, may not understand them, but we know that by our standards they are poor. What we do not know is if they hold the same view when using their standards for judgment. We know that people can be different. This includes holding different standards for judging the world. When looking at people different from us, we must accept that understanding and truth are in the eye of the beholder. Both their world and its truths could be markedly different from ours.

CHAPTER 10

Consequences of the Urban Revolution

This chapter spans three periods of human development. The first is well before any Urban Revolution and any substantial forming of hierarchal societies such as chiefdoms and kingdoms. The second is a transitional period from the end of this early period up to the urbanization of the Urban Revolution. The third is from the start of urbanization to our modern period. In our discussion of the superorganism, we developed a picture of the first period. In this chapter, we are going to skip the middle period and go straight to the third, to the consequence of the Urban Revolution.

We are skipping the middle period because it is transitional and too complex. It is neither "this" nor "that"; it is something in between, and there are too many different examples with which to contend. However, by skipping, we gain clarity. We will be able to contrast two distinct times rather than deal with the blur of societies in transition. Practically, if we did not skip, you would get tired of reading long before I tired of writing, and that would do neither of us any good.

Developmental Periods

We have covered a lot of variable ground to get to this point. It is time to regroup and look at the terrain, and the matrix presented in table 10.1 provides such an overview.

There are four periods in our matrix, with a transition at each juncture. These periods and their transitions do not correspond to specific times, such as 550 CE or 2250 BCE. Populations, just as children, develop at their own pace, where some start to walk at eight months, others much later. Likewise, populations and children both grow in fits and starts. Some populations have yet to make the last transition of the Urban Revolution into the modern period, and even now, some few remain two transitions behind as egalitarian foragers.

With the exception of the transitional period, people and their properties are either this or that; it is binary. The transitional period is different because people are in the process of changing from this to that; it is not binary.

When we first presented the Urban Revolution, we did so using Childe's list of its ten characteristics. On this list are two characteristics with direct applicability to our analysis. One is stratification of society, and the second is the change in identity of the individual from the tribe to the state; we will discuss both in this chapter.

Childe's list of ten characteristics does not a definition make. There is another, one with greater utility for our analysis. It is "the process by which small, kin-based, nonliterate agricultural villages were transformed into large, socially complex, urban societies."[1]

The Urban Revolution was transformative. On one side of the transformation are the superorganismic, "small, kin-based, non-literate agricultural villages" of our story, and on the opposite are "large, socially complex, urban societies." With the pretransition village, there is group selection, whereas on the side with the city, there is individual selection. Individuals on the group selection side identify with the village, whereas on the side with the state, the state itself is the object of the individual's identification. Equality is the hallmark of the old, and the new brings with it inequality

TABLE 10.1   Properties of Developmental Periods

		Developmental Period			
		Chimplike	Egalitarian	Transition	Modern
Property	Civilized	No	No	No	Yes
	Natural selection	Individual	Group	Group to individual	Individual
	Super-organism	No	Yes	Yes to no	No
	Tribal	Yes	Yes	Yes to no	No
	Access to women	Alpha male	Altruism	Altruism & political power	Economic & political power
	Making a living	Foraging	Foraging	Subsistence agriculture	Varied
	Political equality	No	Yes	Yes to no	No
	Economic equality	No	Yes	Yes to no	No
	Status or social equality	No	Yes, except altruists	Not quite as much as before	No
	Social structure	Hierarchal	Egalitarian	Mainly egalitarian	Hierarchal
	Homo	Bellicus	Bellicus	Bellicus to economicus	Economicus
	Individual identity	Group	Group	Primarily group	State

and hierarchy. Finally, those transitioning to the state are civilized, whereas those remaining behind in the village are not.

We have defined those whose identification remains with the pre-state group as uncivilized. By this definition, tribal people living in superorganismic villages are uncivilized, and city people are civilized. The antithesis of the state is tribal organization.

Demise of the Superorganism

Introduction. At some distant time, we were all individuals within our superorganismic groups. Some of these groups still exist; examples are the Yanomamö and the primitive villages in Indonesia we discussed earlier. There is an obvious question: why isn't the superorganism present in modern societies?

Subsistence agriculture. This is a type of farming where the farmers grow enough to feed self and family. It is production for consumption, not sale. It is not commercial. Its sole purpose is to generate sustenance for the producers. It has the property that, once sustenance levels are met, labor ceases. This type of farming continues to exist in rural sub-Saharan Africa, parts of Asia, and areas of South America.

Today's practitioners of subsistence agriculture may or may not be superorganismic groups, but their means of livelihood, if not the same, are very similar. If they are not superorganismic, the unit of production is the extended family and not the multi-family commune of the superorganism. Their only difference is in organization.

The city and the superorganism. Before the city and the Urban Revolution, we lived in superorganismic groups who typically practiced subsistence agriculture. With the creation of cities, they remained in the countryside and continued as before. What changes with the city are individuals leaving the group to earn a livelihood in the city. The city does not eliminate a niche; it creates a new one. As long as the superorganism remains well separated from the urban environment, there is no ambiguity.

The state and the superorganism. The state is something else. It takes ownership of land and controls it for its own purposes. It also conscripts and takes away individuals of the superorganism. Both acts are fatal.

The superorganism requires the availability of its individuals to perform labor in support of the group. This is the only way in which it can function. When the state conscripts for its military or for state projects, this availability ceases. Conscription is just like killing. Conscript enough and there are too few remaining for the superorganism to function. Kill off enough cells of any organism and it will likewise die.

The bigger issue is probably state control of land. The state needs revenue, and in early times, its only source was agricultural production. This requirement competes with the superorganism, which needs land for producing its sustenance. Remove access to land and it dies. The powerful state wins this competition.

We discussed an example in Indonesia. Land owned by others and used for copra or rubber production could not be used by the peasants for their usual subsistence agriculture. They had two choices for survival. They could work for the plantation or move. Not owning the land, they could not practice what was in an earlier time their means for making a living: subsistence agriculture.

The state, in selling or leasing its land to commercial enterprises such as rubber producers, probably believes it is entering into a win–win–win situation. The commercial enterprise earns money, the state receives compensation for the use of land, and the peasants have an opportunity to earn income.

Perhaps the state reached an incorrect conclusion. Why would a peasant prefer working for the plantation, where forty hours a week of probably unpleasant work are required to meet needs? After all, in the good old days of subsistence farming, a peasant could comfortably meet her needs with twenty hours of labor per week and then use the remainder for higher-priority activities, such as socializing.

Coexistence of the state and the superorganism is possible, provided the state taxes the group and not its individuals. This might have occurred, but in the end, the state taxes individuals, and the group as an organism ceases to exist.

There is another issue. The superorganism as a cohesive group, one well honed in the art of exercising unity of action, is a potential threat to the state. The state can manage the miscreant individual; a similar trait in a group is problematic. Multiple groups deciding to take action contrary to the interests of the state could be an existential threat. Consequently, states would be expected to minimize the possibility of such threats by forcing the demise of the superorganism, leaving behind only its individuals.

Once it comes down to only the state and the individual, the group ceases and individual selection takes over. It becomes, as in the very beginning, everyone for herself.

The state is the reason for the superorganism's demise. Where it still exists is probably a testament to the state's weakness, not to any enlightened management. When we later study sub-Saharan Africa, we will see that in difficult rural environments, states have difficulty projecting their power, thus enabling the superorganism to exist.

The preceding scenario has a deeper consequence. With the demise of the superorganism, individuals must totally depend on themselves for survival. Before, with the superorganism, survival was a group effort, and with a group effort, even the weakest and laziest survived. The ethos of no one starves unless all starve results in the smoothing out of the humps and bumps of individual survival. Without the group, individual survival is more tenuous. This forces the sharpening of the individual's evolutionary pencil, as those with lesser adaptive traits are washed out and survival of the fittest returns to being an individual, not group, effort. It becomes very personal.

Consequential Changes

Introduction. Several important changes to our nature resulted from the Urban Revolution. The biggest was the change from group to individual selection. In this section, we are going to discuss that and other changes.

We change identity from group to state. A subject we have mentioned several times in our narrative is changing identity. To now, it has been sort of an ill-defined placeholder, not a well-clothed concept. It is time to dress it up.

Before the Urban Revolution, individuals identified with their group. However, the state results in the demise of the superorganism, and because the superorganism no longer exists, the individual cannot identify with it. However, identifying with groups remains part of our nature, and we can belong to more than one group at a time. The state is one such group, and we normally identify with it.

In the purest sense, we did not change identity; it just appears so. We lost one, the superorganism, and gained a new one, the state. However, the degree and nature of our identification with these two objects are probably different. Does a tribal Iraqi identify with both the tribe and the state? Probably. Does the Iraqi identify with both equally? Probably not. However, the true answer is more nuanced.

Changing identity is more than raising your hand and making a pledge to the new state and its flag. Like ants defending the nest, individuals defending the superorganism will often place the good of the group ahead of the preservation of the self. This is the mark of strong identification.

Does the same degree of identification exist with the state? The response in the United States to the threat presented by World War II suggests that we are as willing to sacrifice for the state as we were for the superorganism. Perhaps the attitudes evolved under one set of conditions carry over when conditions change. The leopard keeps its spots wherever it is, and the degree and nature of our identification with the state might not be different after all. What is different is the object of our identification, and the importance lies with the change, not its object.

A modern example of changing identity. Although it was not explicit in our discussion, we have already discussed changing identity. We earlier related that in Indonesian peasant villages, when there is insufficient land to meet the sustenance needs of the village, some must move away, and moving to a city like Jakarta is an option. When an individual makes such a move from village to city, it is an example of the Urban Revolution. This is evidence that the Urban Revolution is not only of the archaic past but also of the present. It is a revolution still in process.

In the village, the individual is part of the superorganism, and it functions to meet the needs of its individuals. In the city, there is no such entity, and meeting needs is now up to the individual. Consequently, in the city, selection operates directly on the individual, and individual selection is the *modus operandi*.

For the individual to leave the embrace of the village, to become an individual standing apart and free from the superorganism, there had to be an alternative, someplace to earn a living independent of the village. Until the advent of the city, that place did not exist. Survival for the individual was dependent on the superorganism and its associated tribe or village. It was that or nothing.

We lose the cocoon of our superorganism. In our analysis, we have discussed the superorganism as an all-encompassing entity where we, as its individuals, take care of it and it, as our group, takes care of us. It is a symbiotic relationship, just like a bee and its hive or a termite and its nest. They, individual and group, are all of a piece. That the individual belonging to a superorganism strongly identifies with the group should be a given.

The superorganism provided intimate directions for our actions, ones derived from the group's customs and traditions. Losing the group meant losing its direction and becoming dependent on our individual devices. Not only did we lose our room and board; we also lost directions for living. With the transition, it became survival of the fittest individual, a markedly different condition from survival of the fittest group.

There is more to the superorganism than simply room and board and directions for our actions. It is where we satisfied much of our social needs, especially those beyond family. States do not offer much in the way of social nourishment. That is a job for what we call society, a poor replacement for our intimate and enveloping superorganismic groups. The inadequacy of the state and its associated general society in meeting our social needs could be a source of societal ills.

Individuals in the modern, post–Urban Revolution world are not units of a superorganism; they are individuals with agency that is unconstrained by the structure of the superorganism. Society is our new group, and it retains aspects of the superorganism. Society is not the singular, all-encompassing group of our pre–Urban Revolution period, but it still constrains individual behavior. Society can execute for proscribed behavior and shame the line jumper by honking the horn. We may not be part of a superorganism, but we still behave as if we were.

The superorganism is replaced by the state. The state takes over some of the functions provided by the superorganism. A prime example is providing protection from outside forces. However, other functions are not assumed by the state, the biggest being the provisioning of sustenance. Unless there is a welfare system, individuals must fend for themselves to receive their daily bread. Before, the industrious and lazy survived equally; after, one does, and the other, perhaps not.

Natural selection is now individual. Group selection is dependent on the equality of individuals within the group. Recall that selection pressure is a function of the variance of the trait under selection, and with the condition of equality, there is no pressure. Political inequality comes with the state. Economic and social inequality arrives with the city. Consequently, group selection like that with the superorganism disappears. Without the group, all that remains is individual selection. It was always present, but within the context of the superorganism. Now it is all we have; it does it all.

The shackles of equality are removed. The superorganism maintained egalitarianism and thereby shackled us one to the other. No one individual could move very far from the average. We could not be industrious unless all were. We could not improve our status in life unless all did. Individual ambition was constrained by that of the collective. With the superorganism's demise, all those constraining shackles disappear. We can now be all we can be or, on our choice, not.

We made a powerful evolutionary step with the enabling of individual selection concurrent with the freeing from group-focused constraints. Individuals with ambition are free to exercise it and, if it results in adaptive behavior, succeed where other individuals will not.

Individual selection, along with the enabling of inequality, was a great unleashing of evolutionary force. Individuals, not collective groups, made the technical advances that define our modern civilization. Galileo, Pasteur, Marconi, Ford, Edison, Tesla, von Neumann, Jobs, and Gates were not societies, nations, or collectives; they were individuals, ones distinct from the common fold. They were individuals unshackled by the collective.

The state becomes the dominant authority. Before, rules came from the collective, and we had to live with what we as members of the collective agreed to. This system provided feedback, as the collective could change rules that were maladaptive for the group.

Maladaptive or not, rules still had to be followed, or punitive consequences would ensue.

The state is similar to the superorganism in that it also establishes rules for living, and just like before, its rules must be followed. However, with the state, there are two possible flavors, each having a nuanced difference in rule making.

In an authoritarian, top-down state, the makers of the rules have immunity and, other than revolt from below, suffer no direct consequences to bad rules. There is no feedback, and consequently, bad rules can persist.

The second flavor is the democratic state, where rules are proximately top down, just as with any state. However, having voters, rules ultimately become bottom up. This results in a system with feedback for providing corrective changes.

This simple difference plays a role in our larger narrative of why some are rich and others poor. In economic matters, rules play a role. In authoritarian states, economic rules should favor the state's rulers. They should consequently prosper over the general populace, and only a few of the many prosper.

In a democratic state, rules set by the state have a feedback system. This means that rules affecting the economy in a manner detrimental to the voting populace will be changed, and more can prosper.

Nations having economic rules favoring the few over the many will have lesser economic performance than those having rules that enable all to prosper. This sentiment will come into play later when we discuss extractive versus inclusive institutions.

Evolution of Income Inequality

Introduction. Income inequality is a topic in current political discourse. We will later understand that, although currently seen in a negative light, it has been critical in developing our modern nations. Without it, modern nations would never have arrived at their current level of affluence. Importantly, it is the starting point for the process of nations becoming rich, answering part of our initial question. In the following, we are going to discuss its genesis.

History of making a living. In the beginning, during the period of our chimpanzee-like past, we made a living just like current-day chimpanzees do: we foraged. Even when society changed from Boehm's hierarchy in the forest to egalitarian, we maintained this mode of meeting our needs. We were egalitarian hunter-gatherers.

With the Neolithic Revolution, our mode of making a living changed to dirt farming along with some animal husbandry. Doing so required changing our nomadic ways to sedentary ones. This started us down a path away from egalitarianism and toward social, political, and economic inequality. This marks the start of our transitional period, one that leads to its termination, marked by the Urban Revolution and civilization.

Our transition to a sedentary agrarian existence did not necessitate any change to our egalitarian ways. In fact, these ways were highly amenable to our communal farming efforts. This cooperative means of making a living was the source of the rural village, an entity still found in today's developing countries.

Making a living was not something the individual did; it was something the group did. The group or village as the survival machine was the rational actor of economic theory. The individual worked only indirectly for survival, as needs for survival came from the collective.

The individual worked for the good of the group and, owing to custom and tradition, expected no reward for work thus performed. We worked for the group, and in turn, it took care of meeting our needs. When our work was exceptional, the group gave praise as remuneration, and this praise could lead to status.

A new way of making a living, one based on specialization, came with the city. Before, most of us earned a living with farming, and we all performed the same labor. In the city, farming was not an option, and in farming's stead, the city offered a multitude of alternatives. We could be a tax collector for the state, an artisan at a guild, the neighborhood baker, or a middleman merchant for ourselves. The city eliminates a singular means of making a living—farming—but in its place creates many more.

Cities, by their nature, require labor specialties. In our prior, rural existence, we were all farmers and, like all farmers everywhere, had to be able to do all. There are few specialists in the country. With cities came demand for a variety of products and services, each requiring different specialties and skills. In meeting this demand, we became specialists, a big change from our rural, generalist past.

Specialization creates income inequality. Wealth differential existed before the advent of the city. However, it was based on birthright and therefore unearned. We can find examples in chiefdoms with their rulers, nobles, royal families, priests, and similar others. These elites all fed at the public trough, which had as its sole source of sustenance the earned wealth of peasants. With their unearned wealth, the elites lived better than the wealth's creators did.

The foregoing exemplifies unearned economic inequality, one based on birth and family, not diligence and ability. We had to get to the Urban Revolution and the city to have societies with earned economic inequality.

In the city, we worked as specialists. Specialties have different economic values, with higher-value ones resulting in higher economic outcomes in the form of higher incomes. Skilled artisans made more than their helpers did, and merchants made more than unskilled laborers. Along a different bent, smart middlemen made more than not-so-smart ones.

For the first time in our journey, we had earned income inequality. This was not an economic system reserved for just the privileged few. Provided with free and open institutions, all workers in the city's population could participate. Importantly, everyman could achieve wealth based on ability and not birthright, a distinction with a huge difference, as we will soon see.

Wealth, Fitness, and Natural Selection

Introduction. In a broad sense, Darwinian fitness means the ability to survive and reproduce. In a narrower sense, it is a relative measure of the reproductive success of an organism in passing its heritable traits to the next generation via natural selection. Heritable traits include genes and the heritable aspects of culture. In the process of natural selection, traits promoting fitness become selected, and consequently, their prevalence increases in the population.

Wealth increases fitness. Earlier in chapter 5, we discussed two examples of wealth-increasing fitness. One is from a description by Gregory Clark showing that in England for the period 1250–1800, "the richest men had twice as many surviving children at death as the poorest."[2]

There are two possible reasons for this observation. One is that the rich have more children than the poor do and, all else being equal, have more surviving. However, not all else is equal, leading to the basis for the second reason. Richer families enjoy a higher standard of living than poor ones. They have less of a problem with starvation, and their children are less prone to suffering a life-shortening event. This is especially true in the case of disease, as they have living conditions that are more sanitary. Another factor is that the children of the rich are less likely to be participants in war. In general, children of the rich have a greater probability of surviving to the point where they can in turn reproduce. Simply stated, wealth, by improving living conditions, increases fitness.

Clark's data were for pre–Industrial Revolution England. Does wealth still increase fitness in the modern world? Nettle and Pollet say yes: "selection on male wealth in contemporary humans appears to be ubiquitous and substantial in strength."[3]

The foregoing conclusion leads to the basis for a second consequence of wealth: rich men have greater access to women than poor men do. Generally, we prefer as mates those who exhibit fitness. Because wealth enhances fitness, women should prefer as sires of their children men who are rich. This results in rich men having more mating opportunities than poor men do.

There is another facet, one found in societies where women are chattel, or close to it. There, rich men can simply afford more partners. Independent of the case, either buying or selling, more matings result in more progeny, firmly demonstrating a differential reproductive advantage for the rich.

Wealth and natural selection. Natural selection requires variation. It operates on trait variation, not just the trait, making the term *differential* implicit in the concept of fitness. For example, if a trait affects fitness, but all in the population have exactly the same trait, there is no difference and no basis for selection. Therefore there is no evolutionary consequence, and the population does not change. Change only occurs when there is difference or variation within a population for traits affecting fitness.

Wealth can only play a role in human evolution provided that the condition of wealth inequality exists. This condition did not exist in our earlier superorganismic societies. At that time, we were all equally wealthy or poor, eliminating wealth as a player in the process of natural selection.

Inherited wealth inequality. Variation in wealth in the form of wealth inequality arrived with the development of political hierarchies, which first arose in the Neolithic period. These societies essentially established a two-layered social and political system based on hereditary elites. There were the rulers, their families, priests, military generals, chiefs of lesser tribes, and others. However, the vast majority of the people retained their status as units of a superorganism, maintaining their group equality.

Theoretically, traits leading to the initialization of hereditarily based elite status could be objects of selection. An example would be any traits leading to an individual becoming chief. However, this is a one-shot event because, once initialized, the condition of wealth is then due to inheritance and not selectable traits. Inherited status, essentially luck in getting the right parents, is not a selectable trait; it is not even a trait.

Earned wealth inequality. We have discussed that the city by its nature creates a system promoting earned income inequality. Different specialties result in different incomes, making this is a process for creating differential wealth.

From an evolutionary perspective, this process is necessary but not sufficient to effect changes in the population. It obviously can result in wealth inequality, a factor in fitness. However, it only counts provided the requirement for heritable traits is met. If heritable traits such as intelligence and hard work play no role in earned income inequality, there is zero evolutionary consequence and no resultant change in the population. Wealth inequality has a role only if it is a result of inherited traits. Earned wealth inequality usually passes this test.

Natural Selection Changes Populations

Introduction. We have used the term *natural selection,* but without a definition. It is time to focus down, and having a definition is clarifying. The following quotations give a useful picture. The first two are somewhat formal, and the last two give succinct summaries:

> The bare bones of Darwin's theory of evolution by natural selection are elegantly simple. Typically (but not necessarily) there is variation among organisms within a reproducing population. Oftentimes (but not always) this variation is (to some degree) heritable. When this variation is causally connected to differential ability to survive and reproduce, differential reproduction will probably ensue. This last claim is one way of stating the Principle of Natural Selection.[4]

> As a causal theory natural selection locates the causally relevant differences that lead to differential reproduction. These differences are differences in organisms' fitness to their environment. Or, more fully, they are differences in various organismic capacities to survive and reproduce in their environment. When these differences in capacities are heritable, then evolution will (usually) ensue.[5]

> Many biologists define natural selection as differential reproduction of heritable variation.[6]

> Only selection acting on heritable variation will have evolutionary consequences.[7]

Only earned wealth counts. Evolution is a results-based process. At one level, if wealth increases fitness, evolution does not care if the wealth is earned or unearned, but at the next, it does. The issue at hand is whether the created wealth was due to heritable traits or something else. If the wealth has a basis in heritable traits, they are passed along to future generations, and the population changes accordingly. If not, as in the case of luck, being born into a royal family, or similar, there are no evolutionary consequences.

Traits increase in the population. The evolutionary consequence of differential wealth having a basis in heritable traits is that those traits will increase in the population. Over time, there will be an increase in the percentage of individuals possessing traits favorable for generating wealth. The economic intellect of the population will increase accordingly. This is a key point of our narrative, as it explains the rich.

If intelligence plays a role in aiding wealth creation, and if intelligence has heritability, the population's average intelligence will increase. If the propensity to work hard has heritable components, individuals of the population will inherit this trait and work harder. At the group level, if honesty of a population increases wealth, to the degree that honesty has heritable components, it, too, will increase. The individuals of populations become smarter and work harder; populations themselves become socially less corrupt. We will shortly see that all three happened.

More time equals more change. Populations having a longer history of wealth-increasing fitness will economically out-perform those with a shorter one, given a similar starting condition. The nations of Eurasia that started this process some time ago should currently have better economic outcomes than any Johnny-come-lately. They do, and that is a major reason for differences in wealth between nations.

Mobility is a requirement for change. As we discussed in the preceding section, a population has to go through the starting gate of the Urban Revolution before it can commence on this particular journey. However, that step is only necessary; it is not sufficient. This process is a complex machine with several moving parts and has places where sand can get into the mechanism, bringing it to a grinding halt.

Earlier, when we developed the English Process, we stipulated that it required vertical mobility. This discussion is an expansion of the English Process, and the stipulation remains. Our concern is changing the entire population. For this, there cannot be barriers to individuals moving either up or down. What we are looking for are individuals to move up with fitness-enhancing traits and those without such traits to fall out of the population and move down. The genes and culture of the rich must move downward and spread throughout the lower levels. This downward flux of the heritable traits of the rich displacing those of the poor is what lifts the entire boat.

The Hindus of India represent a population without this mobility. They went through the starting gate of the Urban Revolution some millennia ago but never entered into the process we are discussing. Their caste system sets status at birth, and it can only change in the next life, not this one. This freezes mobility, preventing any movement either up or down. Those with fitness-enhancing traits are stuck under a glass ceiling and cannot move up to replace those lesser endowed located above. The lesser endowed have a corresponding glass floor and cannot fall down to a lower, natural level; they too are constrained. The population as an entirety cannot change; the only change allowed is within the strata of a caste.

Not all play this game. Getting to the modern economic world required passing through the Urban Revolution. Many have not. For these populations, wealth as understood in the modern, affluent world is not part of life. As pointed out earlier and as an example, in some Indonesian villages, individuals avoid acquiring wealth because it detracts from the unity of the local group. For these people, wealth is not fitness enhancing and in fact could be fitness decreasing. "These people" exist in more locales than Indonesia; they inhabit, or have until somewhat recently, large swaths of the developing world.

Institutions can be disabling. Institutions are an important topic for understanding our question as to why some nations are rich while others remain poor, and we will discuss them in depth a bit later. However, some of this topic is required for this section, and we will consequently jump a bit ahead and bring it into our analysis.

This is about extractive versus inclusive institutions, as discussed by Acemoglu and Robinson in their important book on development economics, *Why Nations Fail*:

> Inclusive economic institutions . . . are those that allow and encourage participation by the great mass of people in economic activities that makes best use of their talents and skills and that enable individuals to make the choices they wish. To be inclusive, economic

institutions must feature secure private property, an unbiased system
of law, and a provision of public services that provide s a level playing field in which people can exchange and contract; it also must
permit the entry of new businesses and allow people to choose their
careers.[8]

Extractive institutions do little or none of these. They are "designed
to extract incomes and wealth from one subset of society to benefit
another subset."[9]

A prime example of an extractive economy is Zimbabwe with
Robert Mugabe as president. Their economy is two layered, with
a small elite class receiving a disproportionate level of economic
reward, one that is far in excess of their value to the economy. Only
the relatively few ruling elite profit. They control institutions so
that the majority of economic activity results in rewards going primarily to the elites. In contrast, modern Western democracies represent inclusive economies, where all participate and reap rewards
proportional to value of effort.

From the perspective of our story, only the elite in an extractive
economy participate in the evolutionary aspect of wealth generation. They probably get better at "economy," but because they are
very few, there is no corresponding consequence for the population at large. The general population and its associated nation will
not have their wealth-generating traits increased. They will remain
economically inept.

For this reason alone, extractive nations will never improve
their economic status. They need improvement in the applicable
traits of their entire population; they must become different from
what they are.

The economic system must be right. For the process we are discussing to function and effect changes in the population, the right
type of economic system must be in place.

The "right type" is one promoting the creation of differential
wealth or, in the modern political vernacular, income inequality.
Moreover, as we discussed in a prior section, this must be applicable for a major proportion of the population, not just an elite
few. Another requirement is that the reason for differential wealth

be based on heritable traits, not some birthright. The process infers economic mobility.

Economic systems come in four types:

1. traditional: the economy of the superorganism
2. planned: state socialism or communism
3. mixed: a combination of planned and market
4. market: *laissez-faire* capitalism

The traditional or superorganism exemplifies the perfectly wrong type. It is wrong because the creation of differential wealth is proscribed, as equality must be maintained.

The foregoing also applies to the planned economy. Granted, the ruling elite can have differential wealth, but their percentage is way too small to affect the general population.

The market economy in theory meets all the requirements, but in practice, it might not fully exist in all places advertising its existence. States, even enlightened ones, exert forces constraining natural economic processes. The degree of constraint determines the efficacy of the process in changing populations. The less the constraint, the greater the positive change for the population and its heritable properties. Less regulation should result in both greater prosperity and changes in the population toward increased economic intellect.

Economic inequality motivates dishonesty and corruption. Before the Urban Revolution, inequality existed between groups, not individuals. There was no economic inequality, because any economy was internal to the superorganism, with all its members sharing equally. Dishonesty in things economic did not exist, because such behavior would not result in any advantage.

There is a change to this condition when we lose our superorganism and become the only one responsible for self. Lying, cheating, or stealing can now make sense. The baker, in making the loaf a bit smaller but not lowering the price, gains. The shopkeeper with a balance slightly off can likewise gain. Shortchanging in a hurried transaction is also part of this story. So is simple theft.

There is a risk to this behavior. At a minimum, it can result in shame and loss of customers. It can even lead to executions in some ancient societies or loss of the right hand in a supposedly modern one. However, as long as the risk is perceived to be substantially less than the reward, dishonesty can pay off.

When we changed to individual selection and specialization, we created economic inequality. This increased dishonesty both at the levels of both individual and state. Corruption is the name given to dishonesty at the level of the state.

Conclusion. If there are two societies where one has men competing for access to women using economic fitness and the other has warrior fitness as the criterion, the first will have greater economic intellect. The second will have better warriors. In the game of "rich man, poor man," the first will become rich and the second not. However, the first would be wise not to go to war with the second.

The Evolution of Wants

Introduction. In chapter 9's section "Economy and the Superorganism," we stated, "Later, we will discuss the evolution of wants and its consequence, the modern economy." That is what we are going to do now.

Wants for status display. Before we developed any political hierarchy, we were purely egalitarian. That started to change once we acquired chiefs. Initially that was not a big deal, but in time, as we became further organized, we developed associations of groups. Hierarchy increased.

Our first wants probably evolved for the purpose of status displays and did not appear until after we had developed political hierarchy. The tops of the hierarchy, chiefs, big men, and such, displayed their status using fancy headdresses, bigger huts, or such. A function of these status displays was presumably to inform all as to who was boss.

Utilitarian wants. Another class of wants, utilitarian things improving the quality of life, probably soon followed. Elites, being boss and superior in status to their egalitarian brethren, probably learned early in the game that they could live better than the rest. Not only could they get that special piece of meat or nicer chair without any blowback from the lower class but they could make it the rule, suggesting that they could ignore any group shame for their nonegalitarian behavior. They were probably socially distinct from the superorganism.

Changing population requires a majority. Our initial political hierarchy resulted in a two-tiered system. It had some very few hereditary elites at the top, who, with their superior status, could command satisfaction of their needs and any wants from the lower class. They could live as well as the lower-class peasants were able to provide. The vast majority of the rest remained as before: superorganismic, egalitarian, and without wants.

Our interests are with populations and the means by which they change. Specifically, our focus is the changes within a population of their heritable traits due to Darwinian processes. Changing a population normally requires participation by the majority of the population. A society consisting of a few elites, hereditary or not, and the vastly more numerous peasant class is not going to have population-wide changes simply due to the behavior of the few elites. It can only change when the majority is included in the process. We will assume that wants can only effect changes when they are universal within the population.

Wants for the peacock's tail. Our first wants were probably for status display, and the next in progression were those developed by elites for increasing living standards. Satisfying these wants allowed the elites to enjoy a higher standard of living than the general population.

There was an evolutionary consequence to improving living standards. Individuals having wealth in excess of needs, and investing it in increased living standards, provided for their children a higher probability of survival, a topic we have discussed. These individuals had more surviving progeny. This result establishes wealth as a measure of fitness; the rich are fitter.

As we have discussed, the wealthy being fitter has little evolutionary consequence in societies where only a miniscule fraction of the populace is able to achieve that status. However, with the genesis of the city and labor specialization, many more had the opportunity to improve living standards. Everyman, and not just the elites, could achieve increased fitness, assuming sufficient skills, along with a market willing to pay for them.

Males are continuously searching for new and better ways to get mates. Society learns that males having traits to acquire wealth are fitter. Females seek high-fitness males. There was no way to couple fitness-seeking females to high-fitness males until we learned to advertise our fitness. This led to a new class of wants, ones that are not simply wants whose satisfaction proximately enhances living conditions.

Satisfying utilitarian wants that improve living standards increases the probability of our children surviving to reproduce, making those who can satisfy these wants fitter. Evolution takes notice of this connection and establishes wealth as a human peacock's tail. Those displaying such a tail have more and better mating opportunities. The peacock's tail is an expression of fitness. It says, "I can expend all this effort just for show and still have more than enough to survive." It is advertising. Like peacocks, we humans advertise our fitness, and for the same reason. It increases our mating opportunities. Consequently, we display our wealth with jewelry, fancy cars, and expensive clothes. If we do not actually have real wealth, we can sure make it appear as if we do. I recall in the early days of rising cell phone popularity that many in Hong Kong had fake cell phones stuck to their ears. They certainly appeared just as prosperous as those with real ones did.

Wealth has two consequences for our fitness: it makes us fitter by increasing the survival of our progeny; it also makes us fitter by creating more mating opportunities, resulting in more progeny. Importantly, we get those extra mating opportunities by advertising.

There is a caveat to the second case of wealth-increasing fitness. The first case of improved living standards does not have this caveat. For the second case, what is unsaid but implicit is the requirement for differential wealth superiority. You have to be able to display that you are wealthier than your competitors are. This is a competitive game where the score is based on the best relative display, leading to a class of wants without limits. It leads to greed.

Peacock's tails create unlimited wants. Can a peacock's tail be too big or fancy? If bigger or fancier has a better result, that is, more mating opportunities, the answer is no. This I propose is a basis for "unlimited," as in unlimited wants. When it comes to getting our genes into the next generation, there is probably no limit to our effort. This is the bottom line of all bottom lines.

The display of the peacock's tail is constrained by genetically based evolution. Our display of wealth has no such bounds. In our case, culture sets the limits.

Keeping up with the Joneses. Reproduction is about the drive to get our genes into the next generation. Our children have many of our genes, so why wouldn't we want success for those genes too? After all, in a gene-centric world, the genes would. It turns out that we do too. We want our children to have fit mates, and to help in that process, we again advertise.

Debutante parties, the best lawn in the neighborhood, and a large diamond engagement ring are all of a piece. They are the billboards advertising family and the family's ability to satisfy unlimited wants. They are part of a process for getting the fittest mate for our children and thereby getting our genes another step into the future with our children's children. That is what "keeping up with the Joneses" is about.

Wants and happiness. If wants were simply things wanted, satisfying them should make us happy. By that standard, we should find joy in the new car or nice coat, and to some degree, we probably do. However, there is more to happiness than satisfying wants; there is a larger aspect. As we have stated earlier, happiness comes from having more than what our neighbors have. According to Richerson and Boyd, the drivers of wants are pride and envy:

> The rub comes if human wants are substantially comparative, driven by pride and envy, not the satisfaction of wants that do not excite such ethically dubious pleasures and pains. Unfortunately, the evidence from surveys of human happiness suggests that comparative wants are all that one can detect.[10]

A source of envy can be someone having more than we do, suggesting more to wants than just their satisfaction. Having more, like a better car or nicer coat, is prideful and, more importantly, symbolic. It symbolizes our winning the competition. We compete for our genes' sake by demonstrating our capability to satisfy unlimited wants. If we do this better than our competitors do, we win, and the reward for winning is getting our genes across the finish line. Winning is what makes us happy, naturally, and in the process, we get fancier cars and nicer coats.

Unlimited wants make evolutionary sense, and frankly, so does their logical consequence: greed. This sentiment comes with a stipulation. It only makes sense for societies having wealth as a fitness indicator. Not all do. This is because populations are different, and different for the reasons expressed in this chapter.

The consistent theme. In chapter 6, we introduced the Urban Revolution using quotations from both Gregory Clark and Napoleon Chagnon.

Clark argued that competition for reproductive success changed with the Urban Revolution from being based on "war, social intercourse, and social negotiations" to being based on "economic means," and Chagnon argued that "conflicts over means of reproduction . . . shifted to material resources." Chagnon further stated in his argument that "it was only after polygyny became 'expensive' that these conflicts shifted to material resources."

The following quotation is an apropos conclusion to this chapter. In it, Pliny the Elder (b. 23 CE) complains of the cost of women in civilized societies. "By the lowest reckoning, India, Seres, and the Arabian peninsula take from our Empire 100 millions of sesterces every year: that is how much our luxuries and women cost us."[11] Now, we know why.

The basis for reproductive success was changed by those transitioning in the Urban Revolution, but not as suggested in the foregoing quotations. It was not founded on crass commercial activity, although such activity is certainly practiced wherever sex is a commodity. The new basis is subtler and less direct. It is in the advertising of our wealth or our ability to acquire it. It states that we have the wealth needed to ensure the success of our progeny. It states our fitness and thus makes us sexually attractive. Additionally, displaying our peacock's tail expends resources and energy that we could otherwise use for survival. In so doing, we are advertising that we have so much survival potential that we can waste it on such frivolous displays. It trumpets our fitness.

With the Urban Revolution, the image of the caveman dragging a woman by her hair into the cave fades into the past, and in its place rises the image of civilized man advertising with banners of silk and gold. As always, it is all about Women! Women! Women!

Conclusion

This part of our journey started with income inequality, which has as a result a new basis for reproductive success or fitness. In the next chapter, we will learn that this new basis leads to prosperity, intelligence, honesty, and more. Clearly there is more to income inequality than its negative aspect as expressed in the media; it might even be necessary for the way of life most desire.

CHAPTER 11

The Differencing Engine and the Peacock's Tail

This section is about an engine, one we will call the Differencing Engine. However, ours is not some mechanical contraption; rather, it is one principled on Darwinian Theory, resulting in certain populations and their people becoming different from others. This is about the engine that made some populations more prosperous and, for the same reason, more intelligent. It is also possible that it made these same populations less corrupt. The critical point is that societies not having this engine in their evolutionary pathway remained poorer, less intelligent, and more corrupt. In the new economic world created by the Urban Revolution, they remained as they originally were.

This magical engine is not made of pixie dust or fine toad hair. It is the one described by Darwin, made of Mendelian parts and fueled by heritable culture. It is a reason people, especially populations, are different. In this section, we will investigate how our Differencing Engine accomplished all these marvelous ends of prosperity, intelligence, and decency.

The Differencing Engine

We have already discussed our Differencing Engine, just not explicitly as such. It is what we discussed throughout most of the last chapter and, largely, with the English Process back in chapter 5. Now that we have developed and discussed its details, the following outline view should be understandable and tractable:

1. Our Differencing Engine starts with the two primary consequences of the Urban Revolution: the city and the state.

2. The superorganism cannot survive in the city, and the state, by controlling land and individuals, causes its demise.

3. Without the superorganism, there is no longer group selection, only individual. The individual now has sole responsibility for survival.

4. With its demise, we are unshackled from its constraint of enforced equality. We can now exercise individual agency and, unfettered, can become whatever we choose.

5. The city has a need for specialized labor. This forms the basis for earned economic inequality. This is due to higher-skilled occupations paying more than lower-skilled ones.

6. Those earning more improve their living standards.

7. Wealth increases fitness because the rich, with better living standards, have more surviving progeny than the poor do.

8. Additionally, rich men, because they are fit, have greater sexual access.

9. Evolution makes wealth display our peacock's tail, resulting in its use as an attractant for mates.

10. Evolution is competitive. Now, everyman can compete for fitness by competing for wealth. The source of wealth can now be other than birthright; it can come from individual traits.

11. Heritable traits favoring wealth creation increase in the population. These traits come from genes, culture, environment, and their interactions.

12. Among these traits are those promoting intelligence, honesty, and hard work.

13.Because the rich out-reproduce the poor, their progeny flow down the demographic hill to fill in the slots left behind by the less fit poor.

14.Where this Differencing Engine runs, populations become harder working, more intelligent, and less corrupt. They become different in these wealth-creating traits from those populations not engaging this particular engine.

15.Forces external to the engine, especially inharmonious economic systems and extractive institutions, can impede or even stop its functioning.

Prosperity Is a Consequence of Wealth Acting as a Peacock's Tail

Introduction. We want to develop two concepts in this section. The first entails the definition of prosperity. This normally is the satisfaction of needs and wants, but that definition is a bit too simple. Needs and wants have a distribution of characteristics, making some more important for prosperity than others. We will examine this distribution and discuss its implication for how we define prosperity.

The definition we develop will have a role for the second concept. This is the idea that Darwinian unlimited wants, the ones promoting our peacock's tail, have a disproportionate role in prosperity. We will show that unlimited want is a major fuel for the engine of prosperity.

Classification of wants. We are going to define *wants* as consisting of three contiguous classes: Basic, Utilitarian, and Display. Basic are those required for our survival. Food, water, and shelter are in this class. Needs are Basic.

The next class, Utilitarian, is a bit more complex. These are wants having a utilitarian nature. Examples are the extra pair of shoes, a nicer house, and a car. There is nothing fancy or luxurious here, just plain old utilitarian stuff. Satisfying Utilitarian wants increases our standard of living above survival, but not lavishly so.

There can be a blur in distinction between Basic and Utilitarian. What a rich man considers Basic, a poor one may consider Utilitarian. It is all in the eye of the person wanting.

Display is our third class of wants. This is the class associated with our peacock's tail. Satisfying these wants gets us things we use to advertise our fitness. These are our indicators of wealth. Jewelry, the muscle car, and the perpetually perfect manicured lawn are in this class. In the main, any want whose ultimate purpose is to demonstrate a relative well-being superior to others' is in this class.

Remember, our peacock's tail is all about competition, so it is not just the want for our display; it is the want to make our display better than our competitors.' It is the superlative, the biggest, the most expensive, and the finest. If my neighbor has a nice lawn, I want a nicer one. May the man or woman with the better display win!

Motivation is key to classification. Once we meet our basic needs, we can raise our heads, look around, and decide which Utilitarian wants are first to be satisfied. These are probably mainly Utilitarian because, initially, we probably cannot afford much for fancy displays.

Starting the ascent up the economic ladder, our Utilitarian wants probably have some limits. They are not the unlimited wants of economic textbooks, because once you have a reasonably sized house, an adequate larder, and a few pairs of shoes, there is less motivation to acquire more, bigger, or better. Doing so takes effort, and achieving an incremental utility may not be worth it, especially when there are constraints on resources, including time for social activity.

Wanting something for our peacock's tail is a completely different beast. Is this class limitless for all people for all places and time? The answer is no—just consider the very young and old. However, for those in the mating game, and perhaps for those whose close kin are, the answer is probably yes, or nearly so.

We know when our stomachs are full, and when they are, we usually stop eating. This is an example of feedback. This process may be lacking when we exercise our need to display our peacock's tail. A glance at the modern world informs us that there are many who do not know when to stop. McMansions come to mind, and some refer to this behavior as greedy.

Unlimited wants. Is this peacock's tail our only motivation for the Display class of wants? I do not know, but it is clear that it is such a motivator for some people some of the time. Given that our psychology is very complex, there could be other motivators with a consequence of unlimited wants. The main point is that this Darwinian process for propagating our genes is reason for some unlimited wants. It might even explain most, if not all.

Consequence of Darwinian unlimited wants. For those of us not choosing rich parents, we have to work to satisfy needs and wants. If all we had were wants of the Basic and Utilitarian types and could satisfy them with a four-hour workday, would we work more? The rational actor of economic texts would not; we would find different and probably more enjoyable uses for our time. There would be no motivation, Darwinian or otherwise, to acquire excess wealth. We would do no labor for the now nonexistent peacock's tail. Greed might not vanish, but it would certainly be diminished.

There is an interesting and important consequence to an imaginary world of limited wants. In this fictional world, everyone would only work to satisfy Basic (needs) and Utilitarian wants. Figure 11.1 pictures the issue at hand. It shows income use having a Normal distribution where, with increasing income, its use changes from Basic to Utilitarian, and then, finally, to Display. The top curve is the real world with all three classes of wants. The lower curve is for the fictional world of only Basic and Utilitarian wants.

The important point of the figure is the area between the two curves. For our discussion, we will assume that this is the money available for taxation, which is probably reasonable for a progressive tax system. In a progressive tax system, the majority of the state's revenue comes from the wealthy.

Figure 11.1 shows that a disproportionate amount of tax revenue would come from the Darwinian drive for unlimited wants. A consequence of a population having only Basic and Utilitarian wants is a state with limited revenue. With income thus limited, where would the money come from to fill Leviathan's coffers? Specifically, without unlimited wants, especially those for Display,

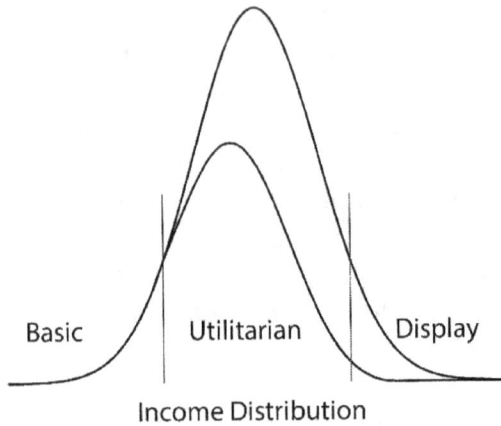

Basic   Utilitarian   Display

Income Distribution

FIGURE 11.1   Classification of Income Use by Wants

driving excessive income, will the state have enough revenue to feed and shelter the poor? After all, they are a constant and always present. The point here is that greed, represented by excessive income wanted for Display, is critical for the modern state to function. Without it, the state would not have the financial wherewithal to perform its function.

What would we do if we actually had this fictional problem? We could let the needy suffer, or we could do as some, but not all, have generally done over human history: we could collectively provide for their livelihood. This effort requires a collective production equal to or greater than the collective requirements for needs and wants. This is what we did when we were superorganismic, and what we do today with institutions like the International Monetary Fund and Médecins Sans Frontières.

That was then, and as the collective, we proscribed selfish behavior. Now, with our state-based societies, we are individuals. We are free of our old collective, and the state permits selfishness. With the state, we work for selves and not others. In this scenario, feeding the poor would require some to work more than required to meet their individual needs and wants. The state would have to coerce this work; else, the poor would die.

State coercion is not required. Darwin came to the rescue by creating unlimited wants in our real world, resulting in unlimited desire for higher income. *Greed* is the word usually ascribed to this condition. In this sense, our quest for reproductive success leads to greed that in turn results in the wherewithal to fund the modern state. Greed is a result of a natural, Darwinian process, which, in the final analysis, enables the state to feed its poor. As Gordon Gekko said in the movies, "greed is good"; it enables the state to feed and shelter the poor.

Prosperity's definition. The foregoing provides a basis for defining prosperity, especially within the context of our narrative. In figure 11.1, assuming no income used for Display, it is the area between the two curves. In a less formal sense, the gauge of prosperity is the amount of wealth used for our peacock's tail; it is income in excess of needs and nominal wants—it is greed.

Implicit in figure 11.1 and in our definition of prosperity is the sense that it applies to society, not the individual. A prosperous society requires enough greedy individuals to satisfy the needs and wants of all. An excess on the right of the curve is required to fill in for the deficit on the left.

The prior paragraph implies a system for redistribution. If the greedy keep all, there will be those whose needs and wants will not be met. Society, in the form of the state, needs the greedy and must take proper care not to kill the proverbial goose through onerous redistribution schemes. To some reasonable extent, the greedy must be allowed latitude to achieve satisfaction. The problem in modern societies is agreeing—to what constitutes "reasonable extent."

The case for inequality. Our Darwinian engine requires wealth inequality. With wealth-increasing reproductive success, as it does, it becomes a measure of fitness, a quality we seek in our mates. This is because mating with fit individuals increases the probability of our genes propagating to the next generation. Individuals, to compete for mates, a scenario where the wealthiest win, use wealth. Competition in evolution is about winning, not just playing the game. The winners' heritable traits propagate, resulting in an increase of those traits in the population.

Winners are those having the greatest excess of wealth over that required for needs and nominal wants. This establishes greed as a trait of winners, and consequently, traits promoting greed should increase in the population. Populations with the higher proportion of greedy individuals should be the most prosperous. This is a not static condition; greed should increase over time, making populations ever more prosperous. This is a feedback system where greed generates further greed.

In a society, Darwinian-fueled wealth inequality is the tide that raises all boats. Assuming a political state that redistributes, excess wealth resulting from greed is fed back to the poor. This system allows all to have their needs met and, if there is sufficient excess, some nominal wants. However, the wealthy will still live better and be reproductively more successful.

We are prosperous because we do not stop working when needs and utilitarian wants are satisfied. We blindly persist without end to maximize the probability of our genes going forward. We appear to lack any Darwinian feedback telling us enough is enough. Would it ever be adaptive to say "enough!" and to cease our persistence? I suggest not, making greed a rational, evolutionary consequence.

In our modern society, greed is a pejorative term. I am not going to argue the case for greed but simply state that it is a basis for economic prosperity. We probably have to accept that "good" prosperity comes with "bad" greed. In our Darwinian world, they go hand in glove.

We did not start our evolutionary journey being capable in matters economic; it is an acquired attribute. Where did we acquire these abilities? They came, like other traits, from genes, culture, and their interactions. They were selected, and for selection, there must be a fitness reward.

Prosperity may not be good. Are those who have yet to transition the Urban Revolution and start this particular journey to prosperity less fit than those who did? I am tempted to answer yes, solely because prosperous people have more surviving progeny.

If there are values having a basis other than Darwinian fitness, the answer is indeterminate. Fitness is not worth or value, it is a biological property like any other. Additionally, fitness is specific for environment, and when that changes, so do the requirements for fitness. What is good for today's world may have no relevance tomorrow, and having made the transition of the Urban Revolution might, in the long term, be deleterious.

Diversity gives strength to populations. Diverse populations have a greater chance of surviving environmental changes, especially sudden ones. With a sudden, large change, the Yanomamö or some similar society could wind up being humanity's sole survivors. This alone is reason enough to value all human life, because, in the final analysis, we want human genes to go forward and continue making footprints in the sand of time.

Increased IQ Is Also a Consequence

Introduction. At the start of this chapter, we stated that our Differencing Engine "made some populations more prosperous and, for the same reason, more intelligent." In the prior section, we discussed how it created prosperity; in this one, we will address its second premise, intelligence. However, this is not intelligence *per se* but rather intelligence specific for creating wealth. It is what we have termed *economic acumen.*

Intelligence is a tricky topic. At the basest of levels, even the experts cannot agree if there is "a single process of general intelligence" or "a multiplicity of different intelligences."[1] Additionally, from a reductionist perspective, we lack understanding of how the brain functions.

Where does our intelligence come from? Some intelligence has a genetic basis. As mentioned earlier, studies of identical twins suggest that about half the variability of intelligence is due to genetic variation. For the remainder of the variance, culture must have a role for no other reason than that it is an important determinant of the environment.

In addition to lacking an agreed-to understanding of intelligence, there are doubts about the metric normally used for its measure, IQ. This is a particularly large issue when between-group comparisons of IQ are made. One issue that commonly arises in this context is the concern that questions asked of one culture have a different cultural context in another.

Even with this uncertainty, we are going to bring IQ into our discussion. We will treat the results of IQ tests as a black box. Our model is that

$$IQ = genes + culture + environment + interactions$$

This is the general model for any human trait. We will assume that IQ measures some aspect of intelligence and, as we will see, correlates with what we have termed economic acumen.

Even though we do not precisely know what intelligence is, nor do we know what the IQ tests are actually measuring, IQ tests produce a number that has strong predictive ability for many of life's outcomes. For example, college students with higher IQs do better in school than those with lower IQs.

Source of data. The following graphs use data for GDP and IQ. The GDP data are the same used earlier in the book, from the World Bank, and are GDP PPP per capita adjusted for natural resources. Nations with populations of fewer than 1 million and those not belonging to one of our eight religions/geographies are not included.

The data for IQ come from *IQ and Global Inequality*[2] (Appendix A), which has ninety-one points in common with the GDP data. The IQ value for each country comes from one to several studies. For example, the data for Central Africa Republic is from a single study of 1,149 adults, and the data for Germany are from nine studies covering 11,162 individuals of various age groups. Where a nation has entries for two IQ studies, the average was used, and for more than two, the median. The data have been adjusted for the Flynn effect, the increase in measured intelligence test scores over the past decades.

Results. In an earlier chapter, we binned the nations of the world into religions/geographies and presented a histogram of the GDP data as "Wealth of Nations" in figure 4.1. It is repeated here.

From that early chapter to now, we have been developing a model to explain the "why" of this graph, the explaination for why

Sub-Saharan
Hindu
Muslim
SE Asian
Latin American
Confucian
Orthodox
Confucian (Less China and Vietnam)
Western

| | 10,000 | 20,000 | 30,000 | 40,000 |

GDP Per Capita

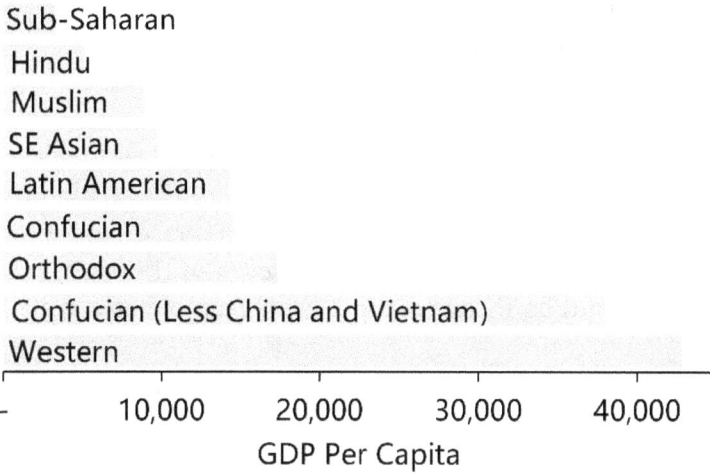

FIGURE 4.1   Wealth of Nations (reproduced from chapter 4)

some are rich and others are poor. We are there, and our answer starts with figure 11.2.[3]

Each data point in figure 11.2 represents the GDP of a nation and its corresponding IQ. A correlation coefficient of 0.84 suggests

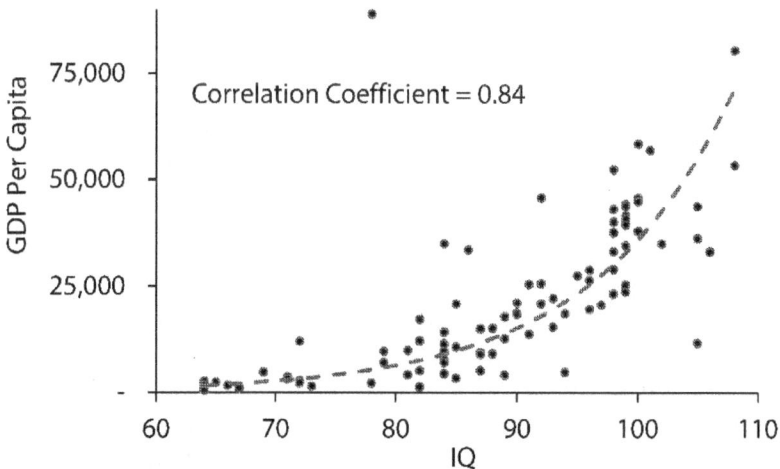

Correlation Coefficient = 0.84

GDP Per Capita

75,000

50,000

25,000

60   70   80   90   100   110

IQ

FIGURE 11.2   IQ and the Wealth of Nations

a very high level of correlation between a nation's economic prosperity and the IQ of its population.

If correlation were causation, this result would suggest that
economic prosperity is the result of smarter populations economically outcompeting those not so smart. For many, if not most, this
would be a reasonable and sensible conclusion. After all, we know
in our own lives that intelligent people tend to make more money
than those of lesser intelligence do. However, in this case, such a
conclusion would be incorrect. This is an example of correlation
not being causation.

From Wikipedia for "correlation does not imply causation," we
have the following:

> For any two correlated events, A and B, the following relationships are
> possible:
>
> A causes B (direct causation);
>
> B causes A (reverse causation);
>
> A and B are consequences of a common cause, but do not cause each
> other;
>
> A causes B and B causes A (bidirectional or cyclic causation);
>
> A causes C which causes B (indirect causation);
>
> There is no connection between A and B; the correlation is a
> coincidence.[4]

Our case is the third one in the preceding list, where A and B are
consequences of a common cause. In our case, A is prosperity, represented by GDP, and B is IQ. Our Differencing Engine is the common cause for both A and B. We have discussed how this engine
causes wealth, and because wealth increases fitness, fitness-promoting traits should increase in the population. Wealth-increasing
traits include IQ. Ergo, prosperity correlates with IQ, but IQ is not
its cause. The cause of both prosperity and IQ is our Differencing
Engine.

Importantly, our model predicts this graph. However, this
statement comes with a caveat. Just because there is a high correlation between GDP and IQ does not mean our model is valid.
Additionally, if this correlation did not exist, our model would be

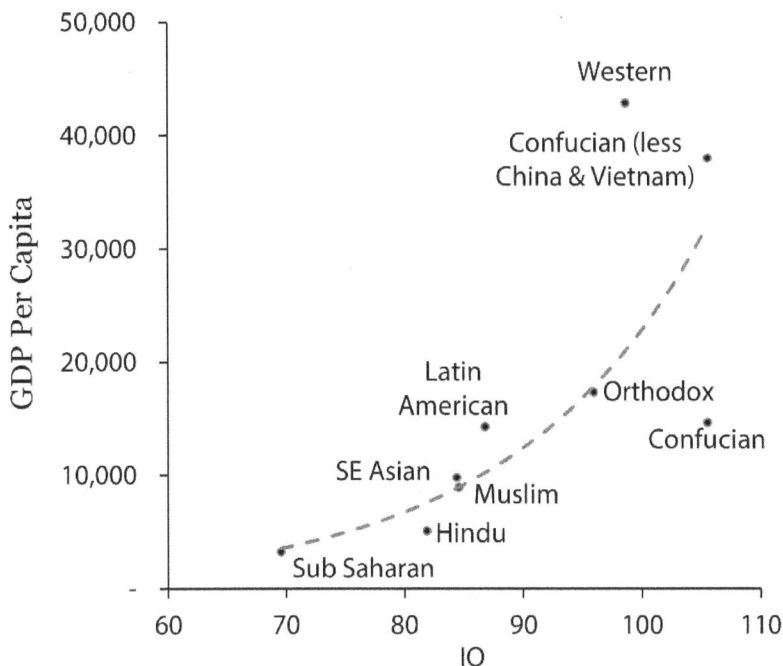

FIGURE 11.3   Wealth and the IQ of Nations

suspect, prompting us to search for our answers down a different avenue. What we can say is that these results are consistent with our hypothesis.

The next graph, "Wealth and the IQ of Nations" (figure 11.3), combines the first two graphs. It does this by binning each of the several nations into one of eight religions/geographies. The values of these data points are not the average of the nations for a given bin; they are the average of the individuals in the binned nations. In other words, it is a weight-adjusted average, with nations having larger populations having a higher weight in the averaging.

Before making this graph, we knew from the "Wealth of Nations" (figure 4.1) that nations differ in their wealth as represented by GDP. We also knew from "IQ and Wealth of Nations" (figure 11.2) that there is a strong correlation between a nation's GDP and IQ. Consequently, the conclusion represented in figure 11.3 is expected. Different nations have different IQs, with those having higher IQs also having a greater GDP.

Similar to actions having unintended consequences, this is an example of an analysis resulting in an unwanted conclusion. The problem is that even though nations are not race, some will make that transpose anyway. When this happens, it makes our conclusion unwanted because we now have to contend with IQ and race, a contentious issue if there ever was one. We can avoid the truth by avoiding eye contact, but truth persists. It will not avoid the reality of our world.

Figure 11.3 shows that some nations are both more intelligent and wealthier than others are. Are we saying that blacks (the sub-Saharans) are intellectually inferior to those from the West or the East? No, what we are saying has nothing to do with race, as in our graphs and analysis, race is not the cause of anything; it is a label without explanatory power. However, ethnicity, to the extent that it defines an evolutionary history, does have power. The cause of differences between nations in both GDP and IQ is the same: our Differencing Engine. To paraphrase the American political commentator James Carville, "it's the process, stupid."

Interestingly, using our analysis with the Differencing Engine as a basis for any comparative statement about wealth would be as equally racist and offensive as one about IQ. In the context of our analysis and our model, "blacks are poorer than those from the West" is as equally racist and offensive as "blacks have lower IQs than those from the West." Such statements become racist only if someone decides to make them so. Even though we might conveniently use race as a player's identifier, the fact that all players have a race does not imply that race has a role. It does not.

We have now answered our original question of why some nations are rich while others remain poor. It is the Differencing Engine, which also results in some having a higher IQ than others. Not unexpectedly, we are left with a new question: why have some nations had the Differencing Engine in their paths and others not? This is for the next chapter.

Data's improper use. It is important to remember that these results apply only to populations, not individuals. We can use our model to predict outcomes for populations, but it has little value for predicting an outcome for any individual.

The Differencing Engine and the IQ of Ashkenazi Jews. In chapter 2, we presented the disproportionate winning of the Nobel Prize by Ashkenazi Jews as an exemplar of "psychic diversity." In this section, we will examine this phenomenon and see how it ties in to our model.

Their having won a hugely disproportionate number of awards in cognitively demanding disciplines such as mathematics, medicine, and economics suggests that Ashkenazi Jews are smarter than most of the rest of us. They are, but not as much as one might expect. The reason for the disproportionate results lies in the nature of the Normal distribution or bell curve, not with the average intelligence of the population.

The average IQ of a European is 100, whereas Ashkenazi Jews have an average IQ of about 75 percent of a standard deviation higher, or 112–115.[5] The difference in winning awards like Nobel Prizes and Turing Awards is in the percentage of the population in the highly intelligent portion, represented by the right-hand side of the Normal distribution curve (see figure 11.4).

Mean IQ = 100
2.3% with IQ > 130

Mean IQ = 110
9.1% with IQ > 130

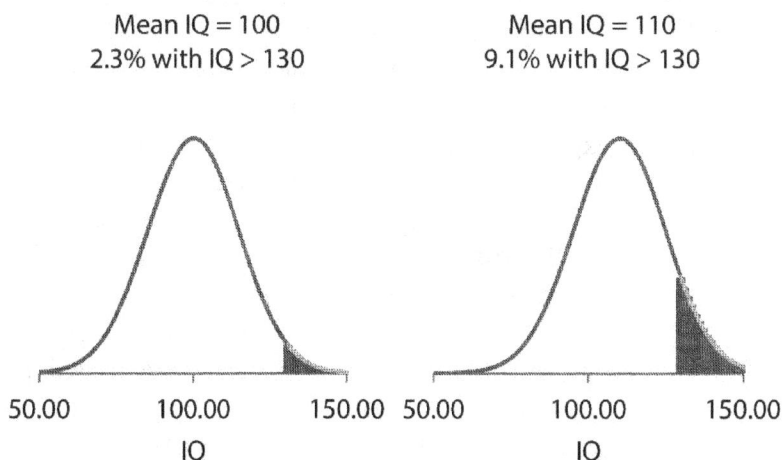

FIGURE 11.4   Normal Distribution of IQ

A population with an average IQ of 100 will have only twenty-three persons per one thousand with an IQ greater than 130. Boost the IQ average by 10 percent, to only 110, and the number of persons with an IQ higher than 130 rises dramatically to ninety-one, or an almost 400 percent increase. On the other side, if we reduce the average IQ to 90, the number per one thousand with an IQ more than 130 drops to a miniscule four.

In our process of answering the question why some nations are richer, we discovered that the explanation offered by the Differencing Engine also explains differential intelligence. Could this engine also explain the high intelligence of Ashkenazi Jews?

There is more to this question. It must be asked and answered within a global context. Sure, the Ashkenazim have high IQs, but when looked at from a global view, so do the Europeans. Theirs is 100, and the five low-performing religions/geographies in our analysis have IQs from 10 to 30 points lower.

Before the Neolithic Revolution, the average IQ of our various populations was probably the same or similar. Afterward, with new, more cogitatively demanding environments, some started to increase and diverge. The IQ of the Ashkenazim probably increased during the same time as the increase in the European IQ, but it increased more. If we assume that the IQ of both the Europeans and Ashkenazim increased for the same reason, our question changes to, what explains the extra increase?

Cochrane and Harpending's book *The 10,000 Year Explosion: How Civilization Accelerated Human Evolution* answers our question. The title expresses the book's central theme, which is the countering of the proposition that human evolution came to a stop when we came out of Africa approximately fifty thousand years ago. Toward its end is a discussion about the intelligence of Ashkenazi Jews. It is a discussion we could "cut and paste" into the middle of this section, as the authors' arguments are almost identical to ours. It would be a wonderful fit, but for brevity's sake, we will not.

The Ashkenazim are the Jews of Central and Eastern Europe, and they first appear in the historical records of the eighth and ninth centuries.[6] Their history is not important to our analysis, but how they made a living is.

They were not farmers but urban dwellers, and as such, they typically engaged in cogitatively demanding occupations such as money lending and trading. Their primary occupation of being moneylenders was prompted by Christendom's proscription against usury, a requisite economic niche they filled.

Successful Jews were prosperous, with a living standard akin to the lower nobility.[7] They also had more children than less successful Jews.[8] Jews as a population had the same evolutionary rules as any population; they might be "chosen," but they are not special. With this differential reproductive success, rich men's genes cascade down to replace poor men's genes. The population becomes richer in the traits that promoted wealth creation. Those traits include intelligence, and the population became smarter.

The foregoing fits our narrative and the Differencing Engine to a tee. Now, let us see why their IQs increased more than Europeans' IQs did.

We are talking about the time before the Industrial Revolution, and during this time, agriculture was not nearly as efficient as it is now. Then, the vast majority, probably 90 percent or more, had to be farmers. This is important from our narrative's perspective, as our Differencing Engine was an urban, not rural, phenomenon. Getting rich and smart meant living and working in the city. Farmers remained unchanged, neither rich nor probably very smart.

Our story is about changing populations. If only 10 percent of a population is subject to change, the results will be proportionate. With the Jews being strictly urbanites, all were players. The entire Jewish population, not just 10 percent or so, could change. This is the first factor explaining the difference.

There is another aspect to the Jewish population. They were endogamous, and through their marriage customs, few from the external population entered theirs. To the extent genes play a role in our Differencing Engine, this is important. Genetic changes cannot effectively occur in a population if outside genes continuously enter. Isolation enables changes to a population so it can become distinct from its neighbors. This is the second factor.

The foregoing does not mean that the population of a nation cannot change. By their being large, genetic changes take longer. It is the admixture of outside genes that mainly dictates the degree or sharpness of change. Stop all outside genes from entering and even a large population can experience significant changes.

Does the Differencing Engine Fight Corruption?

Introduction. At the beginning of this chapter, we stated that the Differencing Engine increased both prosperity and intelligence. We also raised the possibility of a third consequence: it made some populations less corrupt. We have addressed the first two, and it is time to address the third: corruption.

Our framework for this part of our story is a 2012 book, *Why Nations Fail.* The authors are Daron Acemoglu, an economist at MIT, and James A. Robinson, a political scientist and economist at Harvard University. Their thesis is that man-made political and economic institutions are the basic cause underlying a nation's economic performance. They conclude that nations with good economic performance have inclusive institutions, whereas extractive ones are the hallmark of poor nations.

Inclusive versus extractive. The inclusive–extractive dichotomy describes an OFF/ON switch, one that directs the flow of a nation's economy. The inclusive or ON mode enables economic flow to all, and everyone can equally participate. The extractive or OFF mode directs economic flow to only the ruling elite. This results in minimal participation in the economy by the majority nonelite.

A characteristic of the government of an inclusive nation is that it is accountable and responsive to the majority. A major attribute is a rule of law applied equally. In a nation with inclusive institutions, the ordinary individual can "get ahead." Such nations typically have market economies and unbiased laws ensuring private property rights. Individuals are free to open businesses, hire workers, and sell their products and services. Inclusive nations encourage innovation and investment and have rules for maintaining a level playing field. The general population, based on individual choices, participates in the nation's economy. It is a bottom-up economic world, one typified by "We the People," a bottom-up system of governance.

Nations with extractive institutions do not do much or any of this. The government is typically a narrow, elite organizing society functioning for its own benefit at the expense of the masses. Extractive nations often have an authoritative government and, unless they have wealth from natural resources, they are poor. They are top-down economies, with the elite at the top optimizing *their* economic gains via extracting wealth from the lower-downs and handouts from organizations like the World Bank.

Extractive nations dominate the list of the world's poorer countries, especially those in sub-Saharan Africa and the Islamic world. Those in power dole out government jobs to their friends, family, and political allies, a process helping to maintain the power of the ruling elite. Mugabe in Zimbabwe and the late Chávez in Venezuela are examples. The theocracy in Iran maintains power by carving out businesses for its protectorate, the Revolutionary Guard (IRGC). The IRGC is a multi-billion-dollar business empire, and they "are no longer simply a military institution. They are among the country's most important economic actors, controlling an estimated ten percent of the economy, directly and through various subsidiaries."[9]

One of the several examples used in Acemoglu and Robinson's book is North Korea. There probably is no starker a contrast between the economic success of inclusive nations and the economic failure of extractive ones than that offered by North and South Korea. Quoting from the book,

> the economic disaster of North Korea, which led to the starvation of millions, when placed against the South Korean economic success is striking; neither culture nor geography nor ignorance can explain the divergent paths of North and South Korea. We have to look at institutions for answers.[10]

Governments control economic activity through regulation and other activities, such as awarding government contracts to those helping to maintain the status quo. The rulers in extractive nations want to maintain the status quo, so there is no chance of change in the system, one that might result in less wealth flowing to the top.

Tribal nations are extractive, naturally. Two of the religions/ geographies in our analysis, sub-Saharan and Islamic, are mainly tribal, not civilized. As we have discussed, tribal nations are not states in the usual sense. They really are tribes organized in a hierarchal fashion with a statelike veneer covering all.

In these nations, the top man in the dominate tribe runs the nation. Think of Saddam Hussein, Bashar al-Assad, and Muammar Gaddafi in the Islamic religion and Mugabe and Mobutu in sub-Saharan Africa. The top man and his tribe or clan control all the levers of power. The function of the top tribe is to keep power, and for this, brutality is often the tool used.

Mobutu Sese Seko, who ruled Democratic Republic of the Congo from 1965 to 1997, gives evidence with a story related by a family member:

> Mobutu would ask one of us to go to the bank and take out a million. We'd go to an intermediary and tell him to get five million. He would go to the bank with Mobutu's authority, and take out ten. Mobutu got one, and we took the other nine.[11]

Examples do not get much sharper than that.

Natural nepotism. In the world of the selfish gene, our genes want us to take care of our kin. After all, that is where their copies live. Because tribes have a high level of internal kinship or genetic relatedness, it is the nature of nations organized around tribes to be extractive due to nepotism. The dominant tribe takes care of its own, and the out-group tribes remain poor.

As long as such nations remain tribal, I would expect them to remain economically poor. The choice is theirs. They can retain their traditional, tribal ways, but in turn, they should not expect the lifestyles of the rich and famous. This is not a value judgment of tribal societies, just an expression of how the world, Darwinian or otherwise, works.

There is another consequence to maintaining tribal ways. Maintaining civil order in tribal nations requires repressing the inevitable expression of envy by the masses. Remember, the masses want what they see in the ubiquitous media of the Internet. As long as tribally based nations retain their tribal ways and extractive nature, they will remain poor. What the general populace sees they cannot get; envy festers and grows. The maintenance of order in such situations is often accomplished by the populist gambit of shifting blame for economic inequality to the other wealthy, especially the United States and its cohorts.

A force for revolution. The world is not static. The information channels of the Internet along with Internet-enabled social media have enabled as never before the people of the extractive nations to see in real time the rewards of those living in the inclusive world. They want the same. Actually, they probably want a bit more, because as we have discussed, happiness is not having more *per se* but having relatively more than your neighbor has.

A case in point is the Arab Spring of 2012. We can explain some of the turmoil by the usual—groups not in power wanting to be in power. Watching the 24/7 news on cable TV, I came away with the belief that many, especially the twenty- and thirty-somethings, wanted more. I believe that what they wanted is *only an opportunity* to fulfill whatever dreams they have for their future. They innately understood that to achieve that goal, they had to either take over the existing extractive political and economic structures or replace them with new ones. If replaced, we do not know if they would remain extractive, but with change, there is opportunity.

Like those in the inclusive world, they demand to be masters of their individual and collective fates. I believe that the inclusive system they want is not one based on lands too distant. Rather, they want one based on their culture—one they can call their own and of which they can take ownership.

Extractive nations can have economic growth. The explanations are in the book *Why Nations Fail*:

> These tendencies do not imply that extractive economic and political institutions are not inconsistent with economic growth. On the contrary, every elite would, all else being equal, like to encourage as much growth as possible in order to have more to extract. Extractive institutions that have achieved at least a minimal degree of political centralization are often able to generate some amount of growth. What is crucial, however, is that growth under extractive institutions will not be sustained, for two key reasons. First, sustained economic

growth requires innovation, and innovation cannot be decoupled from creative destruction, which replaces the old with the new in the economic realm and also destabilizes established power relations in politics. Because elites dominating extractive institutions fear creative destruction, they will resist it, and any growth that germinates under extractive institutions will be ultimately short lived. Second, the ability of those who dominate extractive institutions to benefit greatly at the expense of the rest of society implies that political power under extractive institutions is highly coveted, making many groups and individuals fight to obtain it. As a consequence, there will be powerful forces pushing societies under extractive institutions toward political instability.[12]

Their thesis and the actual world. History tempers the inclusive versus extractive thesis. In their review of *Why Nations Fail,* Boldrin, Levine, and Modica point out that Germany seems to have done just fine economically under the benign dictatorship of Bismarck, the ruthless one of Hitler, and the liberal democracy they currently have.[13] They further suggest that perhaps it is the just German people themselves who are the reason for Germany's economic success. This is a conclusion consistent with Landes's thesis of culture we discussed early on.

The USSR, a heavyhanded, top-down empire, made impressive economic gains during the first part of the twentieth century. It did so by the forceful reallocation of agrarian resources to industrial production. The USSR failed, and an only somewhat less autocratic and corrupt Russia has risen in its place. It, too, is probably destined for a less than brilliant economic future.

Sure, there are exceptions. Everything has them. However, as we will later discuss, when we add the rule of law to the conversation, the preceding exceptions may not actually be exceptions. In any event, the inclusive–extractive model does a good job of correctly predicting economic outcomes.

Acemoglu and Robinson's model is one of institutions; ours is one of people. They have economic and political theory; we have Darwinian Theory. The models are not mutually exclusive; both can be right. After all, institutions are organizations of people.

Quantifying the problem. In the next chapter, we will attempt to conjoin the two models on a per nation basis. However, we can start the conjoining process now by analyzing a quantification of an aspect of Acemoglu and Robinson's thesis within the context of our model.

Ideas like those presented in *Why Nations Fail* are important, but numbers give ideas dimension and texture. A case in point is corruption, especially as it relates to institutions. Corruption is a mark of a nation with extractive institutions. However, it is important to keep in mind during our following conversation on corruption that corruption is only one part of a larger set of issues contained within the concept of "extractive."

Transparency International is a NGO that receives funding from various governmental agencies (e.g., the Canadian International Fund Agency), companies (e.g., SAP), foundations (e.g., the Bill and Melinda Gates Foundation), and other organizations (e.g., the European Bank for Reconstruction and Development). It defines corruption as the misuse of public power for private benefit. Among other endeavors, it produces the Corruption Perception Index (CPI). The basis for the CPI is expert opinion, not a "quantifiable fact" like GDP.

The graph of GDP and corruption uses the same data set for world GDPs from the World Bank that we have been using. When

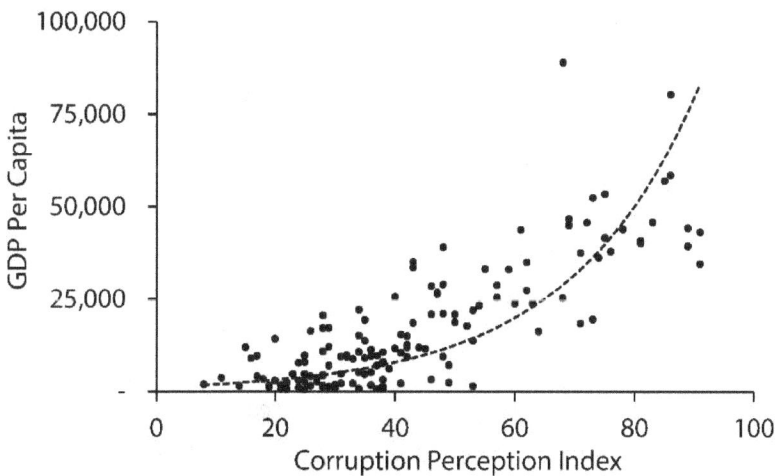

FIGURE 11.5   GDP and Corruption

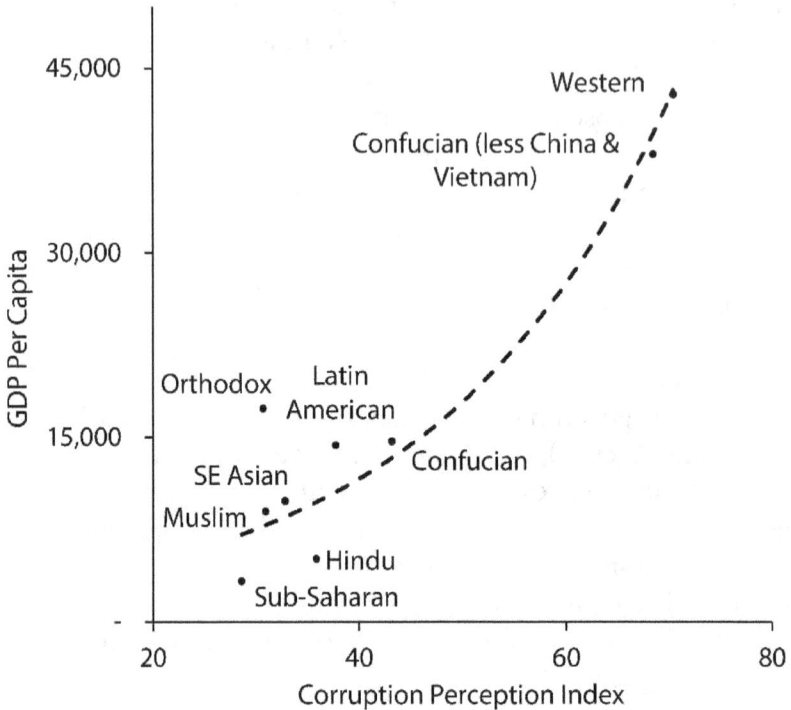

FIGURE 11.6   Corruption and the Wealth of Nations

plotted against CPI,[14] the resulting graph of figure 11.5 is quite similar to the one for GDP and IQ (figure 11.2).

Taking the next step and binning the nations of figure 11.5 into religions/geographies, we get the result of "Corruption and the Wealth of Nations" (figure 11.6).

Does the Differencing Engine result in less corrupt populations? That is certainly a possible conclusion. We can rephrase the question by asking, is corruption's cause the people of Darwinian evolution, or something else? If it were "people," the classes "Confucian" and "Confucian (less China & Vietnam)" should have the same value for CPI, just as they have the same value for IQ used in the preceding graphs. They do not. In the case of CPI, the values are markedly different, suggesting that CPI has a cause other than people and their evolutionary history.

I want to suggest an Occam's razor–type answer, one based on institutions. People are corrupt because they can be. Very simply, corruption exists where legal risk is less than monetary gain. Make and enforce effective anticorruption laws and the problem of corruption should cease.

Acemoglu and Robinson argue that extractive nations are poor, and when we later examine specific groups, we will see that this is true. As discussed, there is more to "extractive" than corruption, but the two are clearly related. With that caveat, we can state that their argument is consistent with the preceding argument, and it is not inconsistent with our model. When we combine the two models, the following conclusion is true: the brightest, most industrious people will remain poor as long as the state allows corruption, especially as part of a larger set of extractive behaviors, to exist on a significant scale. It is important to keep in mind that ridding a nation of corruption or other extractive behaviors does not guarantee prosperity. Prosperity can only come from the people, and extractive behavior, including corruption, only functions as a gatekeeper.

The evolution of inclusive institutions. Where do "good institutions" come from, the ones that are inclusive, not extractive? Acemoglu and Robinson argue that they are contingent on the junctures of history, a view expressed in the following quotations:

> Inclusive economic and political institutions do not emerge by themselves. They are often the outcome of significant conflict between elites resisting economic growth and political change and those wishing to limit the economic and political power of existing elites. Inclusive institutions emerge during critical junctures, such as the Glorious Revolution in England or the foundation of the Jamestown colony in North America, when a series of factors weaken the hold of the elites in power, make their opponents stronger, and create incentives for the formation of a pluralistic society.[15]

> A critical juncture coupled with a broad coalition of those pushing for reform . . . is often necessary. . . . Some luck is key, because history always unfolds in a contingent way.[16]

In the following, I want to argue that contingency, at least in this case, does not hold up as a reason. This will lead to the making of a broader, more important point.

In the fourteenth century, England had a feudal order, with almost everyone engaged in agricultural activity. With the feudal system, lords, who were the landowners, essentially owned their workers, the serfs. The Black Death came along and claimed half of the English population, resulting in a severe labor shortage.

This shortage, like other similar labor–management disputes, gave some advantage to the serfs over their owners, and they consequently demanded change. The English state resisted change. This resulted in the Peasants' Revolt of 1381 and the end of serfdom. With this change, feudal labor services started to dwindle away, and an inclusive labor market started to develop in England.

The proximate cause of change was seemingly the contingent junction of history, the Black Death. In reality, the Black Death was only the determinant of when; it had no say as to what the change would be. That was predetermined because, once power is gained, resulting change comes from demand, pent-up or not, by the group newly in power. In this case, all it took was the condition of serf power over that of the elite. Once that condition was met, the serfs could effect a change to a more inclusive system, one better meeting their wants. The Black Death was only a tipping point; it only determined when, not what.

Institutions' ultimate source. The ultimate cause of institutional change is a more important and general narrative. Once the peasants had advantage, they could force change. The change they wanted and the change they effected increased their fitness. Change was for their good and not the good of any other entity.

Those in power get what they want, and what they want is for the good of their group. If "their group" is "all," resulting changes are on average good for all. This is as it was during the superorganismic days. Good normally ties back to fitness, and because fitness is coupled with wealth, it, too, is part of this simple relationship. Those in power get increased wealth, making political power good for their fitness. In this context, the election cost of a billion dollars or so to become the political head of the most powerful country in the world makes sense. It might not otherwise.

From a general perspective, institutions come from people and their biology. Consequently, they should be created to be good for their creators, especially by increasing the fitness of the creating class. On the reverse side of the coin, people operate institutions, and they would be predicted to operate their respective institutions in a manner that is good for them. In general, what is good or fitter is more money. Wealth grants fitness, so manipulating institutions in both their creation and their operation to provide more money to the operating and creating class is expected.

If society were egalitarian, as it was in the superorganismic times, institutions would enforce equality and operate for the good of the group. Once we divvy up society into groups and associations thereof, institutions should reflect the dominant group's interests. This is because humans look after their own. After all, that is how we were raised when growing up under the tutelage of evolution. Over evolutionary time, we took care of our genes, wherever they happened to be.

If you want inclusive institutions, have everyman develop them, not just some select group. We the people will create inclusive institutions. If you want extractive ones, just leave the task to the elites. You can also get to extractive institutions from inclusive ones by not protecting the majority's power from extraction-seeking elites.

Keep in mind that the elites of today are not those of yesteryear: nobles, lords of the manor, and such. Today they are the moneyed people and their hirelings, the politicians. Moreover, they, along with the politicians' hirelings, the bureaucrats, make up the remainder of the tapestry of power in modern, liberal democracies.

CHAPTER 12

Applying the Model

◇◇◇

We have arrived at this point in our discussion with a model based on the Differencing Engine. It explains much of why some are rich and others remain poor, thus providing the answer to the question posed by this book.

In this chapter, we will address all eight religions/geographies. The main context of the discussion will be the developmental history of each group. We are specifically concerned with how that history relates to our Differencing Engine. We will conclude each discussion with a look at the future.

We have talked about winners, and winners are about competition; so is evolution. After all, "Survival of the Fittest" is evolution's theme song. What is more competitive than the competition for survival and reproduction? Do not make a mistake—as the history of war attests, even at the highest level of civilization, there is competition. At its deepest roots, it has the same basis as two men fighting over a woman—it is all about winning the evolutionary contest.

When we worked with the graphs in the preceding chapter, it was evident that there are only two winners in the worldwide economic sweepstakes: the West, defined by its Protestant and Catholic religions, and the Confucian nations defining the East. Granted, the performance of the Confucians has yet to reach that of the West, but if its primary component, China, continues its ascendency, it will.[1]

Understanding why the West and the East are economically successful is central to our story, and we will spend much of our effort on this topic. As for the poor, they are perhaps more important, but they will not require much discussion. If they have not passed the portal of the Urban Revolution, there is not that much more to say. The story of the winners is deeper and more complex, requiring more effort to understand.

Winners' Interference

Environmental determinism explains why the world's economic winners are only from Eurasia and not Africa, Australia, or the Americas. Offering a greater variety and richness of resources on which to build, Eurasia provided the people of its nations a head start on their journey of development. Environmental determinism enables progression, not its nature. That job is reserved for biology and culture.

It is important to keep in mind that environmental determinism only explains differences in economic results between continents. It offers little in the way of understanding the differences within a continent, especially Eurasia.

Eurasians were the first to arrive at the demarcation point of the Urban Revolution, and if they had only kept to themselves and not wandered from their nominal bounds, their being first would not have been important. Given sufficient time, it is possible, if not probable, that those from other than Eurasia would have followed suit to prosperity, just later. This was not to be, as the winners were not content simply with being first across the line. On arrival in the winner's circle, they strode out into the world, first to explore out of curiosity and then to conquer. Societies on their latent way to states, civilizations, cities, and all the rewarding rest had their trajectories altered.

This interference in the paths of others' development accounts for much of the world's current condition. Without this external interference, in some future century, Africa, India, and all the other colonized lands could have followed the same paths to prosperity as some Eurasians, only later. We will never know.

We can curse the colonizers, and if you are from the West, you get to curse your ancestors. However, keep in mental perspective that, one longer time ago, the West itself was a colony, one of conquering Rome. This is an idea we will expand when we discuss Western Christianity.

The Hybrid Civilization

The concept of transitioning through the portal of the Urban Revolution is a bit more complex than we have discussed. There are two, not just one, components involved. The first is the polity, which sets the governance of society. Before the Urban Revolution, the polity is a tribal, bottom-up organization. After, it changes to a hierarchal, top-down, centralized state. Every civilization making the transition changed its polity from tribal to state.

There is a second component—society—and this is where complexity develops. Pre–Urban Revolution societies were tribal or kinship based, and the individual identified with this societal group. With the Urban Revolution, not all civilizations that changed their polity to the state made the corresponding change in society. They remained effectively socially tribal, with their individuals maintaining identity with the tribe or kinship group. This forms a hybrid civilization, where the polity is a state but the social system remains based on kinship.

We will see later in our discussion that of the Eurasian religions/geographies, only those from the West, the Orthodox, and the Confucians made the complete change. Two of the remaining, the Hindus and Muslims, did not. They both made the change to the state, but socially, they remained tribal, forming hybrid civilizations. This results in a condition where the new polity of the state manages the old society of the tribal collective. The last of the Eurasian religions/geographies on our list, the Southeast Asians, made neither change.

We have defined civilization as the step where the individual changes identity from the group to the state. In the case of the hybrid, change of identity did not occur; it remained with the original tribe or kinship group. This defines those having the hybrid condition of civilization as not civilized, at least by the rules established for our narrative.

Confucians

Introduction. We will start the analysis with the Confucians, but before we do, I want to point out that there is a significant difference between the psychologies of the two economic winners, the East and the West. We could spend a lot of time discussing these differences, but, though interesting, it would detract from our main story.

Nevertheless, there is an example I want to present simply to illuminate the magnitude of this difference.

Amy Chua, in the opening of her book *World on Fire: How Exporting Free Market Democracy Breeds Ethnic Hatred and Global Instability,* discusses the murder of her Filipino Chinese aunt in Manila by her chauffeur:

> For the rest of the family, though, there was an added element of disgrace. For the Chinese, luck is a moral attribute, and a lucky person would never be murdered. Like having a birth defect, or marrying a Filipino, being murdered is shameful.[2]

She continues on, noting that they could not bury her aunt's coffin in the family tomb because doing so would result in more bad luck striking her surviving kin. Owing to the cultural superstition of luck, they buried her in her own tomb, adjacent to but not touching the main one. In eternal rest, she could only be close but never again a physical part of her family.

This story suggests that in the East, luck has properties foreign to the West; it is of a different universe. There are other things, too, existing on both sides of the cultural fence. Those on one side can never clearly see what is on the other; there is cultural obscurity.

Chua's story exemplifies differences between civilizations, and our story is about such differences. However, our concern has been why those differences came into being, and our focus is on those affecting economic prosperity.

Confucian. Confucianism is not a religion but rather a philosophy of the East. It might or might not explain Chua's specific tale, but it goes a long way toward a general understanding of the people from the East.

The Confucian nations of our narrative are China, Korea, Hong Kong, Vietnam, Japan, Taiwan, and Singapore. China alone accounts for about 82 percent of the population, a percentage that becomes even larger when we add in the Chinese-majority nations of Hong Kong, Singapore, and Taiwan.

All seven nations share this common philosophy, but other religions, such as Buddhism, Taoism, and Shintoism, are part of the mix to one degree or another. An individual might worship at various temples of differing stripes because, unlike with the Abrahamic faiths, exclusivity is not required for entry. There is, unlike in other religiously based civilizations, a lack of moral certitude. This results in a *modus operandi* where, when in doubt, you cover all your bases and make offerings at all temples.

Confucius was a philosopher and teacher living during China's Spring and Autumn period (770–476 BCE) of the Eastern Zhou Dynasty and was very nearly a contemporary of Socrates, who lived from 470 to 399 BCE. Confucianism has reigned over the East from that distant past to now, just like the influence of the philosophers of ancient Greece persists in today's West.

Western philosophy has multiple authors, and so does Confucianism. It is not the work of just a single philosopher. There are early Confucians, neo-Confucians, Korean and Japanese Confucians, and New Confucians. Other influences came from philosophers of Taoism, Legalism, Mohism, and others. The East is neither monotheistic nor monophilosophic, but it all started with Confucius, who gets to place his name on the label.

The rules of the Confucian philosophy apply to the state as well as the individual. It includes the Western one of "do unto others as you would have them do unto you." Confucians have the same expression, but in the opposing or negative context: "what you do not wish for yourself, do not do to others."[3]

Confucianism is example, not rule, based. Living the way is far more important than knowing the rule of the way. Actions not only speak louder, they are far more important than words. However, almost contrary to the preceding is the importance of ritual in Confucianism. It teaches that adherence to correct ritual is key to achieving self-mastery. In Western thought, rule and ritual are almost identical; it is not so in the East, where they find distinction.

In the East, the concept of ritual is nuanced:

> Although the subordinate members of a relationship (children to their parents, wives to their husbands) were required to be obedient, their obedience was not absolute and depended upon the superior member of the relationship (parent, husband for example) acting in accordance with his own obligations.[4]

For the purposes of our analysis, the important aspect of Confucianism is in its political teachings, where it establishes a moral basis for governing. The West has its rule of law. China does not, but this does not mean that a vacuum exists. In place of law, China has Confucian moral imperatives that the state is expected to follow. However, when push comes to shove, the Emperor has no constraints. Absolute power is absolute, and contingency of authority in the East replaces predictability of law in the West:

> Confucius advocated for true justice and compassion on the part of the ruler and the ruled. Only by being a just ruler would the ruler enjoy the Mandate of Heaven and continue to have the right to rule.[5]

Consistent with the concept of following ritual is the idea that good governance arrives with everyone playing his assigned role in the hierarchy. "Good government consists in the ruler being a ruler, the minister being a minister, the father being a father, and the son being a son."[6]

Virtue by the ruler is imperative for good governance. "He who governs by means of his virtue is, to use an analogy, like the polestar: it remains in its place while all the lesser stars do homage to it."[7] This leads to the Confucian rationale for not having rule by law, because, in governance, moral virtue is better:

> If the people be led by laws, and uniformity among them be sought by punishments, they will try to escape punishment and have no sense of shame. If they are led by virtue, and uniformity sought among them through the practice of ritual propriety, they will possess a sense of shame and come to you of their own accord.[8]

In the East, accountability is a moral prescription, whereas in the West, it is more certain; there, it is based on procedure. Voting is an example. Unlike in the East, the Western world is black and white. It is based on thou shalt and thou shalt not. It is not relative or contingent. It is not nuanced. The East is relative, contingent, and nuanced. There is a reason the West has a rule of law and China does not.

Confucianism has brought the East a long way, just as Western philosophy with Christianity has brought the West. The West is currently the world's economic leader but, as we will discuss, the East's main component, China, is on its way to becoming the dominant economic power. However, as we will discuss, the East needs the West for its economic future; it cannot succeed without it.

The first step toward prosperity. Becoming civilized is the first step in starting the Differencing Engine. It is a process commencing with state formation. This is when we cast off our shackle of the superorganismic collective and become individuals, individuals identifying with the state. The people who were to become Han Chinese were one of the first to take that step.

Table 12.1, taken from *The Origins of Political Order*,[9] shows the timeline of state formation in China. We have no knowledge of the total number of ethnic groups buried under the number of three thousand polities for the Three Dynasties period, but whatever the answer, back in 2000 BCE, it was probably a lot. Like today's ethnic groups, these pre-state, once independent clans, tribes, and kingdoms probably self-described themselves as proud, unique, superior in all things, and "simply the best," just like ethnic groups do today.

Change, especially of the size and nature we are discussing, takes time. In China, up until the end of the Shang Dynasty, kinship was the basis of organization. The change to state-based polity started with the Western Zhou Dynasty.[10] From there, it took about a millennium and the concluding effort of the Qin Dynasty for a modern state to be forged. When it finally was, the Differencing Engine could proceed. The modern Chinese state was not the only legacy of the Qin Dynasty; it also left for our later wonder the

TABLE 12.1   Timeline of Early Chinese Political History

Year BCE	Dynasty	Period	No. of Polities
5000		Yangshao	
3000		Longshan	
2000	Xia	Three Dynasties	3,000
1500	Shang		1,800
1200	Western Zhou		170
770	Eastern Zhou	Spring and Autumn 740–476	23
		Waring States 475–221	7
221	Qin		1

terra-cotta warriors in the ancient capital of Chang'an, now known as Xian.

The unchanging Chinese family. We have discussed that with the transition of the Urban Revolution, individuals changed identity from the local tribe or clan to the state. In China, this did not happen as advertised. It is as though with the transition, the super-organism shrunk in size from several dozen or even hundreds of individuals to just the extended family. In China, the individual, as understood in the West, never appears, and even today, it remains a stranger. After state creation, China has two entities, the state and the extended family, and not, as happened in the West, the state and the individual.

The foregoing presents an interesting question: why did China wind up with the family, not the individual, as the basic unit of society? For essentially all of China's history, Confucianism has dominated, and this philosophy strongly favors the family over the state. The one time China deviated from this philosophy was during the time of final state formation by the Qin Dynasty.

This other philosophy was Legalism, a philosophy promoting the state over the family. Legalism has a nasty way about it. It provides for "collective responsibility," where the sin of one is paid for by all. This has the smell of divide and conquer, having the potential to destroy the internal bonds of the family. If Confucianism had not replaced Legalism at the end of the Qin Dynasty, China could well have ended up like the West, with the state and the individual as its two entities.

The social structure of pre-state China was centered on the family. It was segmentary, where larger social groups were formed simply by adding together segments or family units. However, the basic segment unit, the family, cannot be divided; it is indivisible. For China, the family unit comprises not just parents and their children; it includes grandparents and other close relatives. It is the extended, not the nuclear, family.

There is a basis other than family for defining a segment. From an operational perspective, it is the smallest group that can fission from the larger group and maintain self-sufficiency. Self-sufficiency in the context of pre-state societies means survival independent of other groups.

The organization of segments is on a basis of lineage, where lineage connotes a kinship group having common descent from an earlier ancestor. Common descent has either a patrilineal (agnatic) or matrilineal thread. Most of the world having segmentary organization was agnatic. Included in this list are China, India, the Middle East, Africa, Oceania, Rome, Greece, and the barbarian tribes that conquered Europe.[11]

The basic nature of such an organizational scheme is that the further back in time one goes, or the more generations that are included, the larger the kinship group becomes. A group going back five generations to a common ancestor is smaller than one going back ten generations. This is the underlying basis of tribal organization. Larger associations of segments are simply made by linking back further in time to an earlier common ancestor.

It is wheels within wheels, and groups within groups, and so on, in an ever-expanding manner until reaching the end at the very first ancestor. Each wheel or group is independent but, for a given hierarchal level of organization, ties back to a single ancestor. Every segmented society has its fountainhead, the theoretical equivalent to Jacob, patriarch of the Israelites. The nature of this type of organization in part explains ancestor worship, common in the East.

The title of this section has its basis in a quotation from Fukuyama. In discussing social development, he writes, "An early stage of development is never fully superseded by later ones."[12] The point is that understanding China is best achieved within the context of an essentially invariant family social unit, a constant of Chinese history. This unit has important properties, ones aiding our understanding of China's society.

Its main property is that it is like a miniature superorganism. For China, it appears to be the unit of natural selection, unlike for the modern West, where it is the individual. With the Chinese family, all its individual units from concubines to oldest son function for its greater good, making it a collective organization. The collective nature of the family defines the collective nature of China and the entirety of the Confucian nations.

As we have discussed, in a Confucian society, everyone has an assigned role. This is true for the family organization. It is a rule-based system, and the rules promote the success of the family enterprise, never the individual.

The family segment is the lowest common denominator; there is no further subdivision. There is no individual, at least as understood in the West. In China, individuals have no importance in the greater society; they are only important within the context of the family. The following quotations from the first prime minister of Singapore, Lee Kuan Yew, help illuminate:

> Eastern societies believe that the individual exists in the context of his family. He is not pristine and separate. The family is part of the extended family, and then friends and the wider society. The ruler or the government does not try to provide for a person what the family best provides. . . .
>
> History in China is of dynasties which have risen and fallen, of the waxing and waning of societies. And through all that turbulence, the family, the extended family, the clan, has provided a kind of survival raft for the individual. Civilizations have collapsed, dynasties have been swept away by conquering hordes, but this life raft enables the civilization to carry on and get to its next phase.[13]

> To us in Asia, an individual is an ant.[14]

The Differencing Engine and the Chinese family. In China, if there is only the family unit and no individual, how does the Differencing Engine function? If individuals can succeed or fail and go up or down the social hierarchy, so can families. There is no reason why they cannot; it is just a bit more complicated. It has more moving parts and thus is less efficient. Being more complicated probably results in greater time lags. Nevertheless, it still should function.

Family mobility has an example with none other than Confucius. Generations before he was born, his family was highly ranked. However, at the time of his birth, his family status had fallen to what we would call middle class. Like individuals, families go up and down, and for the same types of reasons.

The Differencing Engine is a competitive system. In Western societies, competition is normally at the level of individuals. In China, families compete. They are like small companies or enterprises competing in the marketplace. In China, marriage is akin to a corporate merger, where the best merger or union is the one making the respective family units more competitive. Just like individuals and the Differencing Engine, families winning the competition have more of their genes passing into the future. This concept applies to all levels of society, not just the rich.

Mutual economic interests, not affective bonds, hold the Chinese family together.[15] This gives their institution of marriage a basis in enterprise and not love or lust. Obviously, such an impersonal basis for mating can reduce reproductive fitness, a problem the Chinese culture has solved with concubines. As an example, the official number one wife sometimes buys for the family head a concubine. *Buy* is the operative verb, and only the rich can afford this luxury.

This gives the genes of the Chinese rich a double action. With their wealth providing improved living conditions, they get increased survival of progeny, just like in the West. However, with concubines, they get, in addition to more surviving progeny, more progeny. The mistress of the West is the concubine's counterpart but probably has never been as effective in producing offspring. The East wins this battle.

The Chinese family, as a collective enterprise operating over perpetuity, strives to preserve its wealth and uses its inheritance practices to meet this end. There is more than one type of inheritance strategy available, and in China, starting with the Zhou Dynasty, male children split inheritances equally, a practice prevailing into modern times.[16] This equality of male inheritance is an expected practice in the collective family. The family in its entirety is the unit of survival, and the objective is to bring the entire unit forward. Just like for our superorganism, it is all or none.

For China, equality of male inheritance tosses sand into the Differencing Engine's gearbox. The engine requires downward as well as upward mobility and depends on the heritable traits of the rich to replace those of the lower-level poor. When all the males inherit, as in China, all have support, and none falls directly to the bottom, placing a delay into the system. If the family as a unit is incapable of success, it falls down the demographic slope toward the poor families; only successful families remain at the top. Mobility in this instance refers to the family, not the individual. The engine still functions, but at the familial, not individual, level.

The foregoing is a bit different from what we find in preindustrial England. There, just the oldest receive inheritances; all the others must succeed on their own. If the individual does not succeed, he or she falls down the demographic slope toward the level of the poor. The English method results in a far more robust Differencing Engine; it is quicker to get results.

Mobility enablers in China. Patrimonialism is a system where those with power dole out government jobs to family and friends; qualifications are not a subject for discussion. It is a persistent, age-old problem, especially in several modern-day developing countries. The main issue, from our narrative's perspective, is that patrimonialism blocks mobility. Those with heritable traits promoting wealth creation do not advance; only the related and connected do. The Differencing Engine suffers.

Patrimonialism is a battle China fought and eventually won. The dynasty following the Qin, the Han Dynasty (207 BCE–220 CE), started the process of eliminating patrimonialism when it initiated a process where government functionaries were chosen based on merit, not connections, kinship, or otherwise.

In a patrimonially based system, the connected have greater fitness than those not connected, and relatives and friends of the powerful do not easily surrender when Darwinian fitness is involved. Consequently, patrimonialism reasserted itself following the impersonal administration of the Han Dynasty.

Starting with the Tang Dynasty (618–907 CE) the pendulum began its swing back away from patrimonialism. The dynasty instituted an Imperial Examination for selecting functionaries to run the state. Initially, only the elites took the examination, and they still got the jobs, but now they had to be qualified.

The pendulum reaches full apogee in the Ming Dynasty (1368–1662 CE), when anyone, especially others than the usual elite, could enter the competition of the Imperial Examination.[17] Heredity did not count, and everyman could now compete with whatever skills and traits he possessed for the highest grades. Those achieving top grades got well-paying government jobs. Becoming a Mandarin, a bureaucrat of Imperial China, now became primarily dependent on heritable traits, not heritage. China's Differencing Engine is fueled and running.

There is more to China's Differencing Engine than the Imperial Examination. For the engine to reach maximum effectiveness, everyone must be in the game. The space available for bureaucratic wannabes to fill is limited, especially in relation to the totality of the population. The majority of any pre–Industrial Revolution population is the peasant or commoner. In medieval Europe, these were the serfs, and, with their legal bond to their lord and his land, they had constrained mobility. In the West, the decline in serfdom is associated with the Black Death of the mid-fourteenth century. In China, this transition occurred approximately four hundred years earlier, after the Song Dynasty came to power in 960.[18]

We now have two principal components to this Chinese Differencing Engine. One is created by the Imperial Examination and, although open to all, in practicality, only the middle class and above could probably afford the education needed to pass the examination. This first component is robust but has limited size.

The second component, the one for the peasants, is larger, but not nearly as sturdy. Given enough time, both get the job done, separately or together.

When we are finished with our examination of our eight groups, we will see that the one with the longest-running engine is the Confucian. In chapter 15, we compared the IQs of the various groups, and the Confucians came out on top. Does our model explain this result? In totality, it might not, but it certainly offers some explanation. It could be that Confucians have a very high IQ for the understandable reason of the operating time of their Differencing Engine.

Underperforming modern China. In our analysis, we separated China and Vietnam from the other Confucian nations. The rationale was that these two have authoritarian, communist governments, thus limiting their economic potential. The others in the Confucian group have liberal democratic ones, freeing their economic potential and resulting in greater economic success. They represent all they can be; China and Vietnam do not.

The former British colonies of Hong Kong and Singapore have Chinese-dominant populations and do very well in the economic arena. The China spinoff, Taiwan, also does well, as do the non-Chinese Japanese and South Koreans. They are all top-tier economic performers. The common element of this group's success is a combination of democratic governance and market economy.

Taiwan is currently an economic powerhouse with a liberal democratic government having inclusive political and economic institutions. However, it started its rise from the ashes of post–World War II with authoritarian, extractive political institutions. The Taiwan Economic Miracle of rapid industrialization and concomitant economic growth started in the late 1950s, and it was not until 1987 that martial law ended. In the case of Taiwan, inclusive institutions seem to have been a result of economic success, not its cause. South Korea's story is similar.

Taiwan's Confucian cousins, China and Vietnam, might make the same voyage. They both have totalitarian political institutions but partially inclusive economic ones. Communist China was an economic disaster until it opened up part of its economy to free enterprise starting in 1978. However, there is still extensive state control, and many large, important industries, such as steel, are state owned.

In the early 1990s, I visited in China some state-owned enterprises and saw firsthand that a large percentage of the employees just sat about and did absolutely no work. This is the Chinese way of keeping peace with the masses. They get state welfare but have to live at state-owned dormitories and eat at state-run cafeterias, all under the guise of gainful employment. With this charade, the mollified masses allow the state to maintain its power. It is food for peace.

I write this in the summer of 2015. China has economic problems resulting in a major downturn in its stock market, thus prompting the state to intervene by providing artificial support. I suggest that this intervention was to maintain quietude with the newly investing Chinese middle class. Management is fearful of the masses and will do what it takes to prevent even a hint of mass displeasure, which could possibly lead to serious consequences.

In the next few decades, if China persists with state- and not market-driven economic management, it could run up against economic stops, with the consequence of reduced or stagnant economic growth. The alternative is to accept the opening of the political environment to a more inclusive one.

China will never achieve the economic results of the West until it behaves like the West. If the ruling elite insist on hanging on and preventing the requisite changes toward institutions with greater inclusiveness, the masses of billions could yet revolt. This is not just my opinion; notably, Acemoglu and Robinson[19] predict that the Chinese economy will falter unless they make the change.

The risk that is China. The importance of China cannot be overstated. It has a huge population, which is at once both smart, indicated by its top-ranked IQ, and industrious, with its characteristic trait of high need for achievement or N-Ach. From the perspective of potential economic performance, it is unmatched. Sometime in the near future, it will be the biggest economic power in the world and will remain so for the foreseeable future. The rest of the world, especially the West, must adapt to that fact of life.

If China were just another member of the club of the West, there might not be much concern. They are not, and there is concern. China, as it currently exists, is a potential existential threat to the world. Even when China arrives at its destiny as world economic leader, it will still be far short of its potential; it will still have a long way to go to equal Taiwan on a per capita basis. There is a responsibility that comes with being the world's leading economic power; it is the responsibility of stability. A Chinese sneeze will be able to make governments fall and nations fail. In the future, China as the number one economic power will have, given its size, the ability to take down the world economic system. This is indeed scary.

We should expect China, as any other state, to act in its own interests. Furthermore, because it has been autocratic ever since China became China so many years ago, we should not anticipate any significant change toward liberal democracy. What we do not know is if, when China reaches its ascendency, it will change its view from parochial to global. Will it accept the responsibility that comes with status as the world's economic leader? The response to that question is fraught with risk.

China's interest is not that of "China." Rather, it is that of an elite cadre within its ruling Communist Party; they are not necessarily the same. The Party's interest, as for any other organism, is its own survival. This basic interest of the Party will be paramount, superseding what the West might consider best for China itself. In this, the West might not understand, and misunderstanding can lead to bad consequences. Additionally, inopportune decisions by a small, self-interested cadre could have catastrophic consequences for others. Uncertainty of action, along with misguided and unpredictable decisions, is what makes an economically dominant China especially dangerous.

China's future. China will become the world's leading economic power, but it will never become its economic leader. Economic leaders, just like leading companies, require creative destruction. Every now and then, the edifice needs to be destroyed and rebuilt, but rebuilt on the foundation of a new paradigm. How many different foundations have been under IBM since its beginnings as a manufacturer of punched-card data processing equipment (business machines) in the late nineteenth century?

An underlying foundation of the Confucian philosophy is harmony. For example, losing face in East Asia is often associated with inharmonious acts, where expressing anger is one such example. Creative destruction is inharmonious and counter to the Confucian ethos. By their core nature, Confucians will avoid creative destruction, a necessary element for economic growth.

The logical question is, how did they get this far? Easy: they followed. The tip of the economic spear in the world economy is creative destruction, and ever since the Industrial Revolution, the tip of that spear has been in the West. Currently it resides in Santa Clara County, California, the heart of Silicon Valley. Over the past fifty or so years, Silicon Valley has been in the lead in new technologies and business models. The Confucian nations have followed along nicely, making their economic mark by copying and improving on the ideas and methods of others. They have never been the tip of the spear. They might be very good at what they do, but they are not first; after all, it is not part of their Confucian nature to be first.

The reason for the foregoing is in the difference between the collective East and the individualistic West. Being the tip, disruptive, is the job for the strong individual, not the consensus-making collective. Bold new steps are risky, and risk is something the collective avoids. Consider our discussions about the custom and tradition of our earlier egalitarian societies; do what was done before. There is a reason that we all know the names of individuals such as Jobs, Gates, Allen, Ellison, Page, and others; these are the makers of spear tips. Can the collective of Confucian societies create spear tips? I do not know, but, until they actually do, we must assume they cannot.

China may never be a spear tip manufacturer; it is a product of individualists. Unfortunately for China, it is a nation of collectivists. Unless or until China learns to make spear tips, China and the world economy will need the spear tip makers of the West.

India and the Hindus

Introduction. Of the eight groups we have used in our analysis, the Hindus of India are the oldest. Hinduism, while also the dominant religion in Nepal and Sri Lanka, is essentially Indian. In a later section, where we discuss the nations of Southeast Asia, we will see that Hinduism has been influential over an area larger than India proper.

The story of civilization in India starts in the Ganges Plain, the site of the Harappa civilization. Even though this is one of the world's oldest civilizations, by the standards of our story, India remains uncivilized, and in this section, we will explain how this condition is a result of India's caste system.

India's early history. Civilization first appeared in the Indus Valley in what is now Northwest India and Southeast Pakistan. Its term was from 3300 to 1300 BCE, and the name given is Harappa, the name of a village located near its archeological site in Pakistan.

Toward the end of the Harappa civilization in the fifteenth century BCE, Indo-Aryans migrated from Central Asia to India. They brought their Vedic culture, forming the basis of Hinduism.

The caste system. Understanding India requires knowledge of its caste system. First, it is not singular, as there are two distinct but overlapping systems. One, the Varna, has origins in the Vedic culture and Hinduism. The other, the Jati, is of uncertain origin.

The easiest to understand is the Varna. It is a caste system ostensibly based on occupation but actually based on perceived purity and freedom from pollution. The Varna assigns individuals into one of four castes. In order of ranking, they are as follows:

1. Brahmins (priests and teachers)
2. Kshatriyas (rulers and warriors)
3. Vaishyas (farmers, artisans, and merchants)
4. Shudras (laborers and service providers)

The highest, the Brahmins, are pure, and the last, as their lowest, impure counterpoise, are the unranked and unlisted Dalit untouchables. The first three castes represent "second born," which are reincarnated from an earlier life. Shudras are "first born." Those not members of one of these four ranked castes are the untouchables or Dalit.

The critical point in this ranked system is that, because status is set at birth, change can only come in the next life, not this. Nevertheless, even though it is fixed, there is incentive to do well, because doing so in this life increases the chances of being born into a higher status in the next. Unfortunately, for the Hindus, our Differencing Engine only works with mobility in this life; it has no foreknowledge of the next.

The second caste system is the Jati, and it has a nuanced basis for classification. From one perspective, it is the endogamous group into which one is born. Everyone is born into a Jati. It is not just for Hindus, as Indian Muslims and Christians are also born into a Jati.

Properties of Jati

1. A Jati is an endogamous lineage or kinship group.

2. It is reproductively isolated; people only marry within their Jati.

3. After family, a person's primary loyalty is to his Jati.

4. In India, there are more than three thousand Jatis.

5. Similar to the Varna, the Jati system ranks status from high to low. Unlike the Varna, there is no uniform ranking system; each local area has its own.

6. Every Jati has a hereditarily prescribed occupation, a concept similar to the Varna. Like the Varna, each occupation has a relation to purity and pollution, factors used in establishing its rank.

7. However, the occupation of a Jati is not fixed. It can change with the collective will and action of the Jati.

8. There is another basis of classification, one concerned with the complex relations between Jatis. For example, a Jati will accept food and water from only certain Jatis, not all.

9. A village is a collection of mutually interdependent Jatis. This organization of castes enables a broad division of labor. The baker caste needs the plumber caste, which in turn needs a funeral caste when someone dies.

10. A Jati is a closed, self-governing community, with an internal council overseeing its operation. This council has the power to exclude from the group offenders of the rules.

11. An individual's status is that of her Jati; it is fixed at birth. There is no individual mobility.

12. The status of the Jati itself is not fixed. With unity of action, the Jati, as a group, can take steps to improve its status. It can move up by improving its economic status and emulating higher-ranked Jatis. Even though it can move up in the social system, the individual remains fixed to the status of the Jati.

Ancient India left behind few documents, making knowing the origin of the Jati system uncertain. It is possible that it originated with the Hindu religion. The eminent Indian historian Romila Thapar notes another possibility. She suggests that the Jati resulted from a conversion from clan.[20] Support for that view comes from Fukuyama, who states that the Jati caste system was superimposed onto an existing lineage system.[21] If true, this suggests that the modern-day Jati represent the frozen kinship organization of many centuries past, offering an explanation for what we observe in India today.

Whether the Jati is a result of religious practice or simply conversion from clan, is not critical. What is critical is that it freezes individual mobility. No matter his skills or aspirations, the status of the individual is invariant. This social rigidity of the Jati caste system circumscribes the individual's life and living. Importantly, it prevents operation of the Differencing Engine, explaining why India is where it is today: economically poor and relatively low on the IQ scale.

The Indian state and society. Our superorganismic societies were self-sufficient. Everyone could perform all the tasks required for survival, and pointedly, there was little if any specialization. Slash-and-burn agriculturists, like the Yanomamö, are examples.

The Indian caste system breaks that mold. Each Jati, the equivalent of a superorganismic group, has a specialization. This makes the Jati a specialist, just as individuals are specialists in the cities of the modern West. Consequently, survival depends on the presence of other similarly organized Jatis, each having a different specialization. For a village, the aggregation of specializations must be complete to the point that the needs of all are met. There is no self-sufficiency at the level of the Jati.

We do not know the how and why of Jati creation, only that it happened. Assuming that early Indian societies were like other parts of the world, basing society on castes, with each having a job specialty, is a major change from the usual. It is not clear how a superorganismic, generalist society evolves into one that is specialist.

Assuming a starting condition of generalist, superorganismic groups, specialization can only evolve with interaction between specialist groups. For example, a pot-making Jati can only exist if its needs can be met with the several specializations represented by all the other Jatis.

There are two groups in the hierarchy. One is at the level of the specialist Jati itself. The second, the assemblage of Jatis, is the one defining the village as a group. Each Jati is dependent on the entirety of all the Jatis in the village for survival. Perhaps, in India's caste system, the village became the object of natural selection.

Fukuyama, in discussing India, makes the point that religion, not war, formed society. It forged a society bound by an inviolate caste system. "India thus created a strong society and a weak state, the inverse of the Chinese situation in which, then as now, the people have seldom challenged the state-controlled institutions."[22]

Members of several castes constitute a complex village society. Each caste has a council providing self-regulation of the individual Jati. Additionally, all the village castes are linked to each other through a traditionally determined system of barter for services and produce. This makes the collective of Jatis self-regulating and interdependent for survival and thereby creates strong, tightly knit village entities. The caste system makes society strong because it is self-governing and highly interconnected at the group level. Other than for meeting external security and affairs, it has no need for a state. In reality, India's village societies do not need the state; the state needs them. There should be no wonder at the weakness of the Indian state.

The state in India is weak for two reasons. One is the strength of the social system. As discussed, it had little need for the state, thus constraining state power at its onset. The second reason is that in the Varna, priests have a higher rank than rulers do. This prevents the state from obtaining absolute powers, and it cannot freely exercise its will. In India, priests make the rules and kings obey them, constraining the ordering of society by rulers to form a state.

India's history is reflective of its weak state. Before unification and state formation at its independence from the United Kingdom in 1947, India had only a singular episode of unification. This was with the Mauryan Empire, which unified the country from 322 BCE to its decline starting in 232 BCE. Between then and 1947, India remained nonunified, having several differing polities ruling various parts at varying times. At the time of independence from British rule in 1947, there were more than five hundred princely states, none very strong.

State and society in China and India. In comparing India and China, it is evident that there are two structures: the social and the state. The two exist side by side.

China has a strong, authoritarian state but a weak social structure. Its social structure is that of the extended family. Unlike India, in China, there were no priests at the top of the pyramid. There kings ruled and rulers made rules. There were no obnoxious, mumbling priests in the way of ordering society to fit the needs of the state. At the top of the pyramid was absolute authority and power.

India, in contrast, never could develop a strong state. The top dogs were the priests; kings came in second, and priests could interfere with state organization all they wanted.

With the Jati caste system as a backbone, India developed a very strong social structure. Its villages are self-regulating organizations of caste groups having a complex set of rules for their interactions.

The villages can exist apart from the state and, except for external affairs and security, have no need for it. On the reverse, the state has absolute need for the villages.

In the West, the state comprises individuals, whereas in China, the comparable unit is the extended family. India has as its unit the village. The social structure of India is much stronger than the state, making management, even today, a difficult proposition.

India's future. Some millennia ago, when India became Hindu, it baked into its future the nature of its politics, economy, and society. Their future became one with their religion as Hinduism oversaw both this life and the next. Centuries later, the same was to happen when Islam arose in the Middle East. Both religions oversee the entirety of life in all its aspects, and, in both, life and religion are one and inseparable.

Christians were different. Theirs is a religion having a separation between this life and the next. It permits a duality of existence, and what is important is that for this life, free will has been operative. Other than prescribing behavior needed to get into the next life in good shape, it has always left issues for this life to other authorities, namely, the state. Hindus and Muslims do not have the same freedom of movement; their respective "church and state" are unity.

When Indians created their caste system some millennia ago, they predetermined their economic future. The caste system has gotten India this far. Perhaps it did not get here in an affluent condition, but it got here. However, the caste system blocks the evolutionary change that can result from the Differencing Engine, particularly that due to differential fitness. The population remains fixed, it cannot improve, and the Differencing Engine never starts.

Will India change and adapt to the realities of modern economic life? In the rural environment, where most live, the answer is, probably not. The historical constancy of the rural world coupled with the rigidity of social rules has created tremendous inertia against change. The strong village we have discussed is indicative. Revisit rural India in a hundred or even a thousand years, and do not expect to see differences. Why should there be?

As long as the Jati or agnatic lineage exists, change will not come. It is not going to happen. Given the collective nature of the Jati, the best approach is to bud off individuals from its collective and force them to survive independently of the group. This is what happens in the city, and it is there where we can find leverage for change. Cities are the mechanism for escaping the shackles of the collective, and it is there where the old, collective ways should go away.

Cities offer hope; they have less of a historical inertia with which to contend. If Indian society is willing to break with custom and tradition and permit the city environment to work its way, India might change. Cities and the Urban Revolution led the way to modern economic prosperity with others, and it could with India. The rules have not changed, just the people, and they will, if allowed. As long as society remains shackled to the old rules, nothing will change. However, always remain aware that, even if society removes the shackles, change, if it is to happen, will take a very long time.

It is the obligation of the state to assist in that process. Its obligation is to ensure that individuals can survive and, second, succeed. As part of its obligation toward progress, it must ensure a society as meritocratic as possible. Additionally, in providing a survival backstop, it must take precautions to prevent the ascendency of cheaters and free riders.

Southeast Asians

Southeast Asia appears to be just like sub-Saharan Africa. It is poor and undeveloped, and it does not appear until recently to have passed the portal of the Urban Revolution. We will later see that the onus of much of sub-Saharan Africa's lack of economic progress rests squarely on the shoulders of environmental determinism. For many reasons, it is and has been a tough place to make a living. We cannot make the same cause-and-effect association for Southeast Asia.

If we were to try to fit the round peg of Southeast Asia into Diamond's explanation of environmental determinism, all we would find would be square holes. His story line does not appear to work here. By all accounts, or at least those expressed in *Guns, Germs, and Steel,* Southeast Asia should be developed and its people rich. That it is not is a conundrum.

Environmental determinism and Southeast Asia. To understand Southeast Asia, we first have to consider that it has two distinct geomorphologies. There is the "mainland," the part attached to Eurasia containing the nations of Myanmar, Thailand, Laos, Cambodia, and Malaysia. A geographer would include Vietnam and Singapore in this list, but we placed them in the Confucian bin. The second geomorphology, "island" Southeast Asia, consists of the innumerable islands to the south of the "mainland." Here exist the islands of the Philippines and Indonesia along with the island portion of Malaysia.

Unlike its poor cohort, sub-Saharan Africa, Southeast Asia has excellent communications, both internal and external. Jungles, which it has in common, are difficult to traverse, but it makes up for those impediments with a multitude of navigable rivers. These enable communication, especially for trade, well into the interior.

Trade is important for creating civilization, and lying on the trade route between India and China makes international trade a signature civilizing force for the area. It is important to note that even though international trade had high importance, evidence suggests that they had "an extensive and efficient exchange mechanism within the Southeast Asian world prior to any significant trade with Imperial India or China."[23]

The people of Southeast Asia were not geographically isolated like those of the Americas or sub-Saharan Africa. Anyone with a seaworthy boat good enough for coastal navigation could visit— and visit people did. Visitors initially came from China and India, two major civilizations of early times. Later, Arab traders, Christian colonizers, and many others added to the list of visitors.

The heritage of most of Southeast Asia is Hindu and/or Buddhist and originated in India. Chinese Confucianism only traveled as far as Vietnam, making it distinct from its neighbors. Islam is also an import in several areas, but it exists as an overlay onto the earlier Hindu and Buddhist influences and appeared relatively late.

Natural resources like copper ore and oil are both a blessing and a curse. They grant its holders wealth, but this wealth is unearned. Unearned wealth has led those of some nations to feel entitled to such unearned largesse; they do not expend effort for individual initiative or production. Why work when money is free? Importantly for the formation of civilization, why force your subjects to greater taxable production? Why actually work to increase the kingdom?

For the past two millennia, Southeast Asia owned the natural resource of location by its fortuitous position between two huge trading partners, India and China. Traders had to either portage at the Isthmus of Kra or sail through the easily controlled Straits of Malacca or Sunda. The polities controlling those routes of transit could charge for access, and they did. Accordingly, they were rentiers, charging traders for the right of passage.

Southeast Asia was an *entrepôt* for trade, and that was a major role. However, it also added its own goods to the items of international trade. The area possessed exotic woods of particular and useful properties, but its most important trade items were spices, such as cinnamon and pepper. What is important is that Southeast Asians had nothing significant of local manufacture like the pottery of China or the cloths of India. They traded in natural resources and not manufactured goods. All the people had to do was grow rice for local consumption.

Like most of Eurasia, Southeast Asia had as starting resources abundant domesticable animals and plants. For food they had pigs and chickens and for labor water buffalo. They even had elephants, useful for logging or war. Mainly, they had rice, an almost perfect crop. In the tropics, it is grown year round; it can be stored as insurance against future crop failures; it has a high caloric yield measured per hectare. It is efficient in land use in the engineering sense.

The following by the Southeast Asian expert Nicholas Tarling provides a sense of who premodern Southeast Asians were:

> By the early Christian era, Southeast Asia had skilled farmers, musicians, metallurgists, and mariners. Even though they had no written language, no large urban concentrations, and no bureaucratic

"states" of recognizable proportions, they were nevertheless a highly accomplished people.[24]

History of polities of Southeast Asia. Both China and India have been major civilizations for a long time, and trade between the two has a long history. Southeast Asia, with its impenetrable jungles and the impassable Tibetan Plateau, is not on any reasonable land route; trade by sea was far easier. Even then, with the Malay Peninsula, Sumatra, and Java all in the way, the seaways did not offer clear sailing.

The story of sea trade between India and China is much the story of the "island" of Southeast Asia. Initially the early traders sailed to the Isthmus of Kra on the Malay Peninsula and there portaged goods for trade to the other side. It was near this transit point that the Kingdom of Fu-nan arose in the first century CE on the Mekong Delta.[25] It was from that advantageous point that it controlled the trade routes.

By the third century, Fu-nan received visitors from not only India and China but also Persia, reaching its peak in the fourth century.[26] By the sixth century it was in decline as a result of the change in the trade routes from one portaging the Isthmus to one transiting either via the Strait of Malacca between the Malay Peninsula and the island of Sumatra or by the Sunda Straits between Java and Sumatra.[27]

In the sixth century, the Chen-la kingdom superseded the Fu-nan, changing the economy from one based on trade as an *entrepôt* to one based on rice farming. In the eighth century, they too went by the wayside, replaced by the Khmer.[28] Later, the Khmer created the complex known as Angkor Wat.

What was bad for one kingdom, the Fu-nan, was good for the next keeper of the gates of trade. This was the Srivijayan Maritime Empire (670–1025), founded by Malay chiefs and located in the "island" part of Southeast Asia. This was a maritime-based city-state centered on the Musi River in Palembang, Sumatra.[29] It controlled both the Sunda and Malacca straits and remained as gatekeeper until replaced by a sequence of different "island"-based kingdoms starting with East Java (927–1222), a kingdom whose location provided control over the spices of the spice trade.[30] Now, in addition to control of the two straits was control of the source of important spices, the Spice Islands, lying to the east of Java. The Singhasari (1222–1292) and Majapahit (1293–1528) empires in turn superseded East Java.[31] Eventually the Southeast Asian Maritime Realm (c. 1500) took over.[32]

On taking over the area once controlled by the Fu-nan and Chen-la kingdoms, the Khmer started their march toward creation of their empire. This culminated in the complex of monumental temples and other religious buildings at Angkor in what is now modern-day Cambodia. This city and its surrounding environs eventually covered nearly three thousand square kilometers, becoming "the most extensive urban complex of the preindustrial world."[33]

The economy at Angkor was not founded on international trade; it depended on rice.[34] Rice needs water; water requires management, and Angkor was all about controlling water. It is a maze of canals and dams.

A large urban area and many people do not a state make. States need structure, or at least, important ones do. Structure means institutions, and it is permanent institutions lending the state permanence. "Cult of personality" is not an institution, and that is what the rulers of the Khmer Empire depended on. Tarling states that "orderly conditions and glorious deeds were more the result of their personal abilities than of an institutionalized command system."[35] Additionally, "the pattern of Angkorean history reveals an unstable and seemingly irremediable reliance upon personality."[36]

Both the empire and the city reached their peaks in the eleventh to thirteenth centuries. At that time, Angkor supported a population of approximately 1 million people,[37] a level not reached in Paris until about 1850. Not all civilizations persist, and by the fifteenth century, Angkor and the Khmer Empire had collapsed. Four centuries later, when Raffles was founding Singapore and Anna teaching the scions of the King of Siam, the reclaiming jungle hid what civilization made.

During its six-hundred-year history, and just as they were in other parts of Southeast Asia, Hinduism and Buddhism were important, and religious preference depended on the current king. The temple complex Angkor Wat expresses the ambivalent nature of religion in the Khmer Empire. Initially constructed as a Hindu temple in the early twelfth century, it was later appropriated by Buddhism. By the late thirteenth century, Buddhism was Angkor's sole religion.

In its day, the Khmer Empire was substantial. In addition to Cambodia, it ruled over what are currently Laos, Thailand, and southern Vietnam. It had foreign relations with Java to the south, China to the north, and India to the west. Importantly, it was on the trade routes between China and the Arab and Indian lands to the west.

Several factors are associated with the empire's decline. One was the shift to Buddhism where the religious culture changed from priestly Hindus to monastic Buddhists. Another is the change from an agrarian, rice-based economy to one focused on trade and commerce where they went from a continental empire to maritime *entrepôt*,[38] the seemingly default economy of the region. The third was pressure from their neighbors, the then ascendant Thais. The Khmer abandoned Angkor as their center in the 1430s,[39] long before the arrival of the European colonizers.

In addition to the Khmer Empire, there were two other empires of significance in the region. In the Irrawaddy basin appeared an inland agrarian polity centered in Pagan in present-day Myanmar. It appeared in the ninth century, peaking in 1173–1211. It is the site of monumental temples dedicated to Theravada Buddhism. Buddhist monks can be powerful, and to achieve legitimacy, kings had to cede land to them. Eventually running out of land, they lost power, and the kingdom collapsed in 1281.[40]

Located between Pagan and Angkor, Ayutthaya, in modern-day Thailand, became the third player in mainland Southeast Asia. It was a Siamese kingdom existing from the fourteenth to the eighteenth century, and like Pagan, it was a Buddhist state. In the sixteenth century, foreign traders described it as one of the biggest and wealthiest cities in the East.

With the exception of Ayutthaya, what the European colonizers encountered in the sixteenth century was populations that at one time were under a civilizing influence. They were no more, and since the collapse of the Khmer Empire, they had not been since the fifteenth century.

Why did Southeast Asia remain tribal? Southeast Asia did not become prosperous like the West, and we want to understand why. The first and most obvious answer comes from Tarling, who stated, "Much of the region, even into modern times, has been occupied by peoples who are basically tribal,"[41] and we know that remaining tribal prevents the Differencing Engine. Additionally, adding to the tribal issue is that "the notion that a 'state' was a permanent structure controlled by a 'government' to which obedience was automatically due was not shared by many of the societies."[42] The people were not only tribal; their ethos rejected the concept of state.

These are observations, and although important, they are not fundamental explanations. Perhaps we should consider the nature of the jungle itself. Not only is it is a principal feature of Southeast Asia but it is also found in abundance in other uncivilized or tribal areas, namely, much of sub-Saharan Africa, the Amazon basin, and parts of India.

States and civilization suggest a connectedness of multiple groups of people, and for them to form, communication that enables commonness must exist. The corollary is that factors hindering communication impede formation of state and civilization. The following by Tarling suggests that communication in densely forested Southeast Asia was lacking:

The extensive river basins of the mainland and Java may seem conducive to human settlement, but villages were often separated by wide stretches of forest and by hilly ranges, so that few people travelled regularly outside their own district.[43]

Southeast Asia presented a complex geography such that it "served to encourage the growth of communities which were physically distanced from each other."[44] Connectedness between groups in the same river basin was probably relatively simple, but between groups occupying different basins, it was nearly impossible. People lived on land near water, where the water was usable for travel; land lacking this amenity probably remained unoccupied.

The foregoing suggests that it must have been difficult for a polity to control more than a single watershed. However, with a layered structure, a single polity could control other lower-level polities, each with its own internal communication. Tarling also points out that "a type of political entity had evolved which enabled several kingdoms to join together and yet maintain a fundamentally equal status."[45] However, this is not what states need to form; they need a single entity to control all from top to bottom, not an organization of multiple subentities. All the individuals must identify with a single point at the top.

Southeast Asia pre-1750 had low population density, and according to Tarling,

> the most important factor inhibiting population growth . . . was probably the instability of residence brought about by warfare and raiding, voluntary and forced migration, the pioneering of new cultivation areas, and corvée obligations.[46]

Additionally, "most of the 20 to 30 million people of Southeast Asia were concentrated in a dozen trading cities and in the wet-rice agriculture,"[47] with those outside these areas practicing shifting agriculture. With the exception of rice cultivation, these are conditions similar to what were found in sub-Saharan Africa.

Kings and others in power could control the relatively few areas of significant population density. Outside of these areas, people were scattered and control was difficult to establish. In both Southeast Asia and sub-Saharan Africa, if people did not like the existing authority, they could move.[48] After all, there was plenty of unoccupied land.

People in power demand economic rewards from those lower in the hierarchy. In the West, control of land resulted in economic return. There labor for this return came in the form of serfs legally attached to it. Southeast Asia was different; there "rulers perceived their power in terms of human rather than territorial resources, their object in war was always to capture as many of the enemy as possible, to take home to populate their dominions."[49] A similar situation occurred in sub-Saharan Africa. In both, the ability of the ultimate producers, the peasants, to move away made controlling land of little value; value came from owning the producers themselves.

In this context, slavery makes economic sense, and both sub-Saharan Africa and Southeast Asia had slavery as an economic basis. This is exemplified by the fact that in 1673, in Batavia (Jakarta), the population was 27,068, comprising a grand mixture of ethnicities, but half (13,278) were slaves.[50] Additionally, "slave raiding was another source of inter-communal conflicts."[51]

We now see some of the reasons why Southeast Asia remained tribal and uncivilized. A main issue appears to be the jungle geography, a factor of environmental determinism. However, its effect was not direct but rather higher order. The jungle kept population densities low, forcing rulers to capture people, not land, to achieve their ends. In addition, with low density, people who did not like their rulers could simply move.

When rulers capture people, slavery, an institution inimical to our Differencing Engine, ensues, and when people are able to escape such rulers, they create migrant populations. Moreover, such populations do not form the urban centers critical for the state. If this model is correct, then neither civilization nor the state will form until population densities within a circumscribed area are sufficiently high. As we have discussed, jungles have a role by constraining density.

We started this section with the following:

> By all accounts, or at least those expressed in *Guns, Germs, and Steel*, Southeast Asia should be developed and its people rich. That it is not is a conundrum.

That conundrum might still exist; however, considering the foregoing, it is diminished. Even with Angkor and other similar civilizing polities, Southeast Asia remained tribal. Some might argue that Angkor and perhaps a few other polities were in fact states, but that, in the end, is not important. If they were, they did not persist, and their populations, by our standards, did not become civilized. The question that remains is why, and "jungle" is probably part, if not most, of the answer.

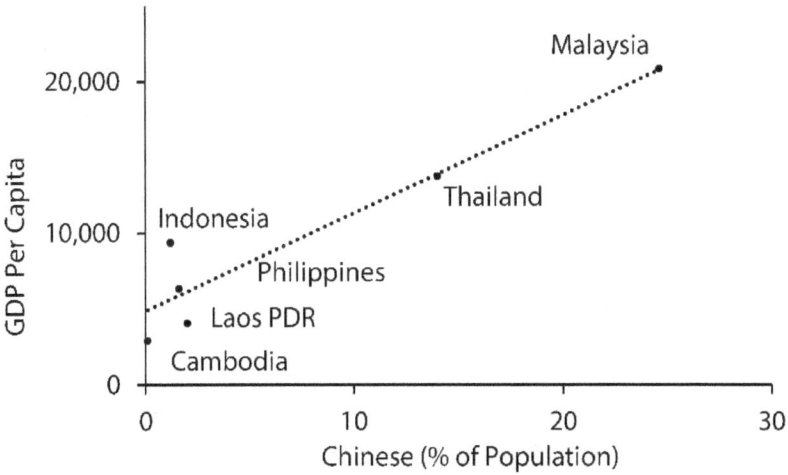

FIGURE 12.1   Effect of Chinese Population on GDP

Lying statistics. The histogram "Wealth of Nations" first presented in figure 4.1 shows Southeast Asia with a GDP of $9,783 and sub-Saharan Africa's GDP at $3,255. We have made the case that the underlying reasons for the poor economic conditions of each were approximately the same. With Southeast Asia's economic performance at three times sub-Saharan Africa's, something is amiss.

The graph "Effect of Chinese Population on GDP" (figure 12.1) explains. To varying degrees, the nations of Southeast Asia have resident populations of overseas Chinese, and they have been a presence since before the start of colonial times. What is clear from the graph is that their presence results in improved economic performance. The intersection of the trend line with the *y*-axis shows that at 0 percent Chinese population, the GDP would be $4,855, a value consistent with our narrative. This is also consistent with the economic performance of China's neighbors, the Hindus, having a GDP of $5,091.

The greater questions and the future of Southeast Asia. The graph raises two questions. Should we expect the presence of an economically adroit population, the Chinese, as a minority group to increase the economic condition of an economically lesser host country? Is figure 12.1 expected? The second question is related to the first: does the rising tide of Chinese economic performance raise the standard of living boat for all? Does the average Malay live better for having overseas Chinese as part of Malaysia's economy?

We are going to reserve the next chapter for discussing comparative economic performances of mixed populations. The second question relates to the economic future of Southeast Asia. We can reframe the question: will an ascendant China influence the economic outcomes of its neighbors? Will the Southeast Asians be better off economically for having China as a neighbor?

The answer must be yes. As China grows and fills in, it will be less inclined to perform less economically taxing jobs. Furthermore, China's labor costs will increase, and when China becomes less competitive, someone must take up the slack. If Southeast Asia lacked an overseas Chinese population, who would fill in would be up for grabs. However, with ethnic Chinese present in the neighborhood, the nations of Southeast Asia should benefit.

I would expect onerous labor to be done by Southeast Asians and the management of that labor by resident Chinese. The nations of Southeast Asia will become similar to America's neighbor to the south, Mexico.

This has important geopolitical implications. China will rule not only China but also the entire set of southern tier nations. It will be a huge hegemon and will control all of East Asia. The East will then include Southeast Asia.

There is a problem with the scenario of the future sketched here: there is a long history of overseas Chinese and Southeast Asians, much of it unpleasant. In general, dominant minorities are ill received by local hosts, and the overseas Chinese in Southeast Asia are a prime example. Their history has at times been bloody and murderous.

Both Thailand and Malaysia have recognized the problem of overseas Chinese as dominant minorities and have taken steps to minimize them. Going forward, Indonesia and the Philippines will have to take similar steps, and so will the smaller others. If they do not, there could be one of two consequences. The obvious is being left behind. However, China will get its way, and if Southeast Asians remain antagonistic to overseas Chinese, China could intervene.

Clearly, answering the question of whether Southeast Asians are better off economically for having China as a neighbor is complicated.

Muslims

Building a state requires getting the individual to be responsive to the state and not the tribe.[52] For millennia before civilization and the advent of the first state, the individual identified with a local kin group. With so much social inertia, effecting such a change is difficult. In this section, we will see that the Muslims used a unique approach to solving this problem. Unfortunately, they were unsuccessful, and consequently, today they remain in an uncivilized condition.

We will use the Ottoman Empire for this part of our story. This empire started in 1299 in the western part of what are now modern Turkey and the Balkans. Over the centuries, its territories greatly expanded to include much of the modern Islamic world along with some of the Christian European, meeting its demise in World War I. We do not need more detail than these starting and ending points and the fact that it expanded its territory.

Before the Industrial Revolution of the eighteenth century, the world was Malthusian. Lack of significant ability to increase food production along with disease kept population growth low. Consequently, there was not much excess capacity for taxation, and any polity seeking to increase its wealth had only one recourse: conquer more land. This is exactly what the Ottoman and other empires did.

Conquering requires armies, creating a need for soldiers, but there are limits to any local pool of potential soldiers. There is of course the physical limit based on pure body count, but there is a second. Using a local pool has the risk of creating an army with divided loyalties, with part of the division biased toward a local power structure or elite and not the state.

The state wants to avoid any competition for ruling authority by local, internal groups and avoids actions allowing any purchase toward such ends. To avoid such problems, conquerors in the process of empire building used the conquered as a source of soldiers.

Conquerors own conquered people, making them slaves. It probably is not as absolute as stated, but the fact is the conquerors used the conquered as a source of soldiers. One day a soldier to be is part of a tribe and willing to defend it. On the day after losing to the conquering, the soldier is part of a unit of the state, its army. The state provides the soldier sustenance, perhaps a woman slave or two for good soldiering, and something new to identify with: the state. As a soldier-slave, the new soldier owes his life and future entirely to his owner and master.

The state forces a change of identity. It does this by enslaving some of the people of conquered tribes and bringing them into the state's fold. The slaves become soldiers, laborers, house servants, and other lower-class occupations, or, as in early China, even human sacrifices to accompany their dead masters in the afterlife. In time, some skilled and talented slaves rise up the hierarchy to serve their masters in higher callings. Eventually, some become free and citizens of the state. The main point is that after forceful separation from their tribe, the conquered or their progeny over time become citizens of the state. The conquered become civilized.

The Ottomans were just like any other greedy group; they wanted more lucre. However, they had an obstacle to forming a conquering army. The Ottomans were Islamic, and in the main, so were the people they conquered. The problem they had was that it is against the tenets of Islam to enslave a Muslim. Consequently, they had to find another source of slave-soldiers.

In lands controlled by Muslims, there are non-Muslims, especially Christians. It was in the Christian enclaves of the empire, especially the Balkans, where the Ottomans found their solution. They took young Christian boys from their families in the form of a levy. The boys were enslaved, then converted to Islam, and, finally, specifically raised to be soldiers for the conquering empire.

This cruel solution was not new to the Ottoman Empire. It was a practice dating back to the Abbasid caliphs of the ninth century[53] and continued into the eighteenth century. For almost a millennium, a visceral hate for Muslims rightfully developed within the Balkan Christian community.

The impetus for the original slave-soldier of the Abbasid rulers was that they could not depend on tribally sourced armies to hold on to empire.[54] They needed troops loyal to the ruler. Conquered, non-Muslim slaves trained and educated for their mission in life did quite well. Separated from family and tribe, they developed two loyalties. One was to their master, the state, and the other was to one another. Having soldiers loyal to one another is good for the successful prosecution of conquest.

Initially, the Abbasids bought their slaves, but in time, a more direct process evolved, one in which it became a form of required levy to the ruler. Periodically, agents of the Caliph visited the Christian villages, identified the most promising boys, and took them directly from their parents. They were never to see or hear from them again. The arrogance and absolute certainty of religious rightness have no limit. In any event, the Islamic state obtained a slave-soldier army, one absent of ethnic or native Muslims.

The slave-soldier was not simply a basic foot soldier. The first step in his development was conversion to Islam. The second was military training, but this was more than basic training. The entire army from top commander to bottom foot soldier consisted of slave-soldiers. Training was for making not only good foot soldiers but also, over time, military leaders.

The Ottomans took the slave-soldier concept one step further. They educated the best and brightest to be civil servants, and just like in China, the best of the best became its Mandarins. Unlike in China, the state's managers were not local natives; they were converted Christian foreigners. The smart ones ran the empire and those lesser fought for it. Taking slaves to be soldiers was not new, but the Ottomans' making them the Mandarins of the Islamic state was. In the end, the Ottomans wound up with a military and civil service loyal to the state.

This was a meritorious system, but one with a catch. The best and brightest could achieve wealth and power. Some even achieved the title of Grand Vizier, prime minister to the Ottoman sultan.

Whatever they gained was not theirs to keep or pass on. If they performed well, they could live well, but upon their end, any acquired titles, lands, or property reverted to the state.

Some could marry, others not. Some were even required to be celibate. The main point is that they could never establish an estate, prohibiting the initiation of any lineage. Keep in mind that no matter how well they performed in the civil service or military, they remained as slaves, and they could be executed at will, for any or no reason.

Patrimonialism has been a problem in developing and maintaining a strong economy. Any state having a system where the good jobs and favors go to friends and relatives of those in power can be expected to have a poor economy. It is an example of misallocation of resources. The Ottoman system of Christian slaves avoided some of this. The slaves, which became rich and powerful, had no relatives to help. This was only a partial solution, as the friends and relatives of the elite class of Muslims could still exercise patrimonial activities.

Having non-Muslim foreigners run the state had the consequence of excluding native Muslims from aspiring to well-paying government jobs. This was obviously a ploy to keep competitors for the top job of ruler at bay. Keeping potential Muslim competitors away from positions of power within the state reduced their ability to gain purchase via being a member of the ruling elite.

Where is the state? There is a native ruling family or tribe. There is an army and bureaucrats for managing the state, but they are foreigners. Besides the ruling clique, no Muslim could be part of the state, either its army or bureaucracy. They remained as members of their tribe and were never true citizens of the state. Given this curious means of state formation, the Islamic state is an oxymoron. It is a collection of tribes with an overseeing central authority behaving as if it were a true state. The Muslim majority remains tribal, and by the definition we are using, they remain uncivilized.

Modern-day Iraq, Syria, Libya, and many others of the Islamic world are inheritors of the Ottoman Empire. They are states in name only. They are organizations of tribes with the head of the strongest one as ruler. For a Bashar al-Assad or Muammar Gaddafi or Saddam Hussein to maintain his dominance as "head of state," he must use the very heavy thumb of the state, as recent history has shown.

Not all the modern states within the boundaries of the old Ottoman Empire are tribal. Some, such as Iran and Turkey, are probably states within the context of our story. They probably developed their properties of state before the advent of the Ottoman Empire.

For our narrative's perspective, the critical issue is that with the Ottoman Empire there is no Differencing Engine; it does not exist with the tribe. The people of that former empire remain poor and with a relatively low IQ.

Future of the Islamic nations. The nations of sub-Saharan Africa are relatively easy to understand. Until recently, they have remained tribal, and this explains much of their current economic condition. Additionally, there is no major religion confounding the issues. With other poor groups, such as the Hindus, religion is a confounding issue. In this sense, the Islamic nations should be similar, but they are not. The reason is Islam.

Hinduism and Buddhism are mainly philosophies for living life. They have their mystical aspects, but dogmatic they are not. Their adherents can believe and practice other faiths. They are not absolutist.

Christians are similar to Muslims, but with a main difference: Christians keep separate the secular and the sacred. Recall the admonition "render unto Caesar that which is Caesar's." They recognize the authority of state as an entity separate from faith. They are not absolutists. Muslims have only the sacred; there is no secular. Everything is under Islamic rule. They are absolutist and, as absolutists, lack tolerance.

Islam, at its founding, started in tribal areas. In the Muslim world, the tribe is part of Islam and there is no separation. Islam and any associated polity are a unity.

There is a consequence to being a unity of faith and polity. When considering any change to improve economic performance, even a simple one, Islam is in the center of the discussion. This intertwining of faith with every aspect of life impedes change, and for this reason, the Islamic nations will remain as they are. Clearly, if economic and political systems do not change, neither will people, and real change means people change.

If we were not all interconnected, this nature of the Islamic nations would not matter. They could lead their lives and the rest of the world theirs. We are interconnected, even to the extent that the Egyptian taxi driver or Omani nurse can listen and see in real time events and people in the non-Islamic world. The Internet has taken away all our clothes and we cannot hide from one another. This is a problem.

The people of the Muslim world can now see how the other lives. Before, this knowledge came via state-controlled media. If Muslims believe theirs is the perfect religion, and they do, how can they accommodate the reality of what they see to their religion's promise of perfection? For this, the affluent West is an existential threat. It is a reality counter to their belief. Their belief and the reality they see cannot coexist. One must be annihilated, just as matter and antimatter cannot coexist in the same space. This threat is amplified when sexually repressed young Muslim men see how young men just like themselves live in other parts of the world.

There is more. A significant number of Muslim nations receive what wealth they have from natural resources, namely, oil and gas. Oil and gas reserves are finite and will someday start to decrease in output. When this occurs, unless they have replacement economies in place and operating, even the oil-rich nations will become poor.

If Islam is to survive, it must accommodate the reality of non-Muslims. This is the truth underlying Huntington's *The Clash of Civilizations and the Remaking of World Order.* Their not accommodating the existence of others is very worrisome.

Western Christians

Introduction. When we shortly finish our narrative, we will see that, based on the narrowly defined standards we have set, there have been only two civilizations: the East and the West. The Confucian East in China started its civilizing process before developing writing, so our ability to gain an understanding of what happened is limited. Western civilization's history is more recent, and consequently, we know more about its genesis.

The Roman Empire was the source of two of our eight groups. Its western half, the Western Roman Empire, has responsibility for the West itself and consists of two major religions: Protestant and Catholic. The eastern half, the Byzantine Empire, with its capital in Constantinople, is the source for the Orthodox religion.

The founding religion of the West was the Roman Catholic Church, and that is where we will start. However, the sociologist Max Weber associated the development of capitalism in Europe with Protestantism. His thesis is that the Protestant work ethic was a major factor in Western Europe's economic development. Consequently, we are going to divide the West into two parts, Catholic and Protestant. Once we have covered the Catholic, we will move on to the Reformation and Protestantism.

We will attempt to keep our discussion general and not specific to any given area in Europe. However, where specificity lends value, we will use Great Britain.

Empire's conquest. Before their conquest by the Roman Empire, the people of Great Britain, like most others in Europe, were illiterate and tribal. At the start of their conquest in 43 CE, what was later to become England consisted of approximately thirty Celtic tribes. This is a number similar to the number of polities in China's Spring and Autumn period more than five hundred years earlier.

Upon conquest, Great Britain's indigenous people became subjects of the empire and became its colony. This is just as India became a colony of England seventeen hundred years later. Although many eventually gained Roman citizenship, the natives were not citizens of an English state. That took until 1066 and the Norman Conquest at Hastings.

Colonizers leave marks on their colonies, and Rome left hers. The first mark was peace. When an empire, such as the Roman, conquers barbaric tribes, peace among the conquered follows. Intertribal conflict reduces the economic value of the conquered, and so the conquering enforce internal peace. *Pax Romana* is the specific term, *Pax Imperium* the general.

Why is there a polity? Why are there states, tribes, or kingdoms? Why was there feudalism, and why did we form superorganismic groups? All these questions have a common answer, one associated with the peaceful consequence of *Pax Romana*. They all exist for the protection of individuals within from individuals without. They are for external security. However, complete security cannot exist until there is also internal security. Only after securing the issue of internal security can the conquering state address other matters. Consequently, the first item on the conquerors' to-do list is internal peace.

The Roman Empire, by enforcing peace within its conquered territories, reduced a principal need for the preexisting tribal organization. When that occurs, the individual becomes dependent on the state for security and no longer requires the tribe to satisfy this need. Lessening dependence reduces the tribe's value, resulting in less incentive for the individual to maintain tribal identification.

A second mark of the colonizing empire was hierarchy. Tribes are egalitarian, and an empire's state brings hierarchy. There are now rich people with higher living standards living among the poorer, egalitarian conquered. Do people and their societies instantly become hierarchal? Of course not, but when the conquered poor view the conquering rich, there must be pressure to aspire to their superior condition. Envy is motive.

A third mark is the city. The Roman Empire was the personification of the Urban Revolution. As Wickham observes, "the Empire was in a sense a union of all its cities."[55]

The Roman Empire brought with conquest the "urban" in "Urban Revolution." Before, there were no cities or towns in Great Britain; London and all others appeared after. Notably, throughout its entirety, the empire created a thousand cities.[56]

Cities are not just buildings and higher population densities. The contrasting rural is a constant where all live the same egalitarian, agrarian lives; there is little change on the farm. The urban environment is opposite. It is hierarchal with a variety of job specializations. For this reason, it is a force driving people to become different. People change, and the city is where this differentiation occurs. Most importantly, it is mainly in the city and not in the rural, agrarian environment where the Differencing Engine runs.

The state, as an employer, is a fourth mark. Before conquest, almost all worked the land for sustenance. The conquering state brought an alternative. Work for the state and the state will provide food, shelter, and clothing or their equivalent in money. Those taken or accepted by the state to be its laborers, soldiers, and bureaucrats start the civilization process by shifting identification from their old tribes to their new source of sustenance and security. State employees, as a percentage of the population, were probably few, but for these select few, the state satisfied their primary needs of security and sustenance. Their changing identity to the state should be a given.

There is a second-order effect caused by the state taking over the security function from the tribe. With tribal societies, when an enemy attacks, all in the tribe muster for the common defense; it is part of the social contract individuals have with their tribe. States are political, not social, and cannot provide security in this manner. They must have a standing army.

The Roman Empire had an army of five hundred thousand[57] comprising mainly non-Romans, or "barbarians." This army, along with the usual administrative needs of the state, required money, and it used taxation to achieve this end. As with any similar state, taxes come from the producers, and in Roman times, these were the peasants. In the good old tribal days, once family and tribe were fed, the peasant could relax and spend time on social activities. The state, with obligatory taxation, reduces the availability of such leisure time. To the extent that the social activities enabled by leisure periods are necessary for tribal maintenance, taxation weakens the tribe.

The two most important elements for survival are sustenance and security. At one earlier time, the tribe supplied both, and consequently, the individual identified with the tribe. The marks of state we just now discussed, those of peace, inequality, the city, state employment, and taxation for security, all have a role in security and sustenance. They all have a role in influencing the change of identity from the old provider, the tribe, to the new one of state.

Empire's fall. Changing human nature takes time, and time is something the Roman Empire did not have. After the Western Roman Empire collapsed in the fifth century, the Dark Ages followed. The empire had simply ran out of time to complete the civilizing process it had started.

Once the impetus of the civilizing process, the Roman state, vanished, the process ceased, and not being sufficiently far enough along for self-maintenance, civilization ended. The evidence lies with the ensuing deurbanization, where cities, the backbone of civilization, essentially ceased to exist. When they are gone, so is civilization. If there had not been a second, parallel state in the Church, Western civilization certainly would not have happened the way it did.

We do not need to detail the empire's fall; all we need are two observations. One is that by the latter part of the fifth century, London was essentially abandoned. The other is similar. Rome, the seat of the former empire itself, went from a population of 1.6 million in the second century to being temporarily uninhabited four centuries later with the sack of Rome in 546[58] (see figure 12.2).

We know that the West became civilized, but how was the process completed without the urbanizing force of empire and state? There are two parts to answering this question. One, as stated, relates to urbanization, a process that, although under way, ceased with the fall of the empire. The other involves the process of changing identity from kinship group to state, where, in this case, the state was not the usual suspect—a king or similar—but rather the Church.

The Church as the cause of Western civilization. We have stated that the reason for the polity of the state is to provide for the common defense. This is exactly what the Roman Empire did, but only for this life and not the next. Those having a belief in life after death need just as much security for getting to the next life as surviving this one. That job was the province of the new institution on the block, the Roman Catholic Church.

Before Christianity, paganism dominated the religious scene; it was the religion of the early Roman Empire. In 380 CE, the Roman Empire abandoned paganism and took to its breast the Catholic

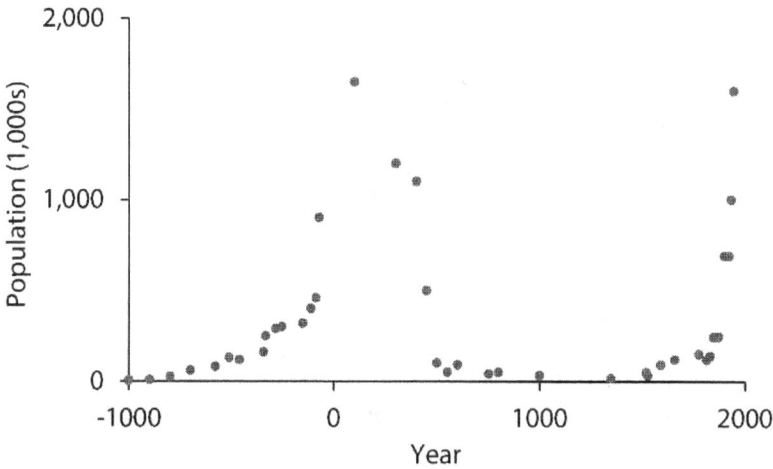

FIGURE 12.2   History of Rome's Population

Church as its official religion. In a broad, nonpejorative sense, the Church was a parasite of the state, as wherever the empire was, so, too, was the privileged Church.

At the fall of the Roman Empire, there were two parallel states. One was the normal political state, the one providing protection in this life for the here and now. The second state was the Church. Like the political state, it was hierarchal and taxed the population to provide its services. After all, cardinals, bishops, and parish priests need food and shelter.

Each state served the same population but addressed different needs. With two states, the individual could be both a Roman citizen and a Catholic parishioner. When identity of the individual changed from the tribe, it could be to state, Church, or both.

Whatever civilizing process existed with the Roman Empire, it should have ceased when it fell. This did not happen, because the parallel state of the Church maintained some level of process continuity. In a world where kings, states, and empires could, and did, change, the Pope and his state were a constant. This established the Catholic Church as an important factor in the creation of Western civilization.

This unique role of the Church differentiates what was to become Western civilization from others. In the others, the change from tribalism to the state was probably coercive. Fukuyama points out in comparing China, India, and the Middle East at the time of the Prophet Muhamad to Europe that "Europe was very different from these other societies insofar as its exit from tribalism was not imposed by rulers from the top down but came about on a social level through rules mandated by the Catholic Church. In Europe alone, state-level institutions did not have to be built on top of tribally organized ones."[59]

This suggests that in the West, state-level institutions were built on those initially formed by the Church, providing consistency between the institutions of the state and the Church.

We have discussed mechanisms for how the state causes change of identity. The Church also effected this change, but via different mechanisms. Tribes are organized around agnatic linkages, and they are dependent on inheritance for maintenance. Keeping property within the lineage is what helps to keep it together, and any rules restricting inheritance of property within the lineage weakens the tribe. For inheritance practices, the Church was an interested observer. It was effectively the inheritor of last resort and thus materially benefited from Christians dying without heirs. Marriage rules are an important part of any system of inheritance. With the Church establishing rules for marriage, it could mold them to its own end, and it did. Fukuyama states, "The church systematically cut off all available avenues that families had for passing down property to descendants."[60]

The levirate system is an example. This was a somewhat common practice in tribal societies according to which a widow is required to marry her late husband's brother. Doing so keeps property in the lineage. By making such marriages difficult, the Church enhanced its chances for acquiring this property. Just as the superorganism managed fitness to promote its interests, the Church managed marriage rules. The difference is that the superorganism got altruists, whereas the Church only gained money.

That the Church was highly successful in this endeavor is evidenced in the following quotation:

> By the end of the 7th century, one-third of the productive land in France was in ecclesiastical hands; between the eighth and ninth centuries; church holdings in Northern France, the German lands, and Italy doubled. . . .
>
> The German, Norse, Magyar, and Slavic tribes saw their kinship structures dissolve within two or three generations of their conversion to Christianity.[61]

By acquiring land that had been in tribal hands for centuries, the Church weakened the tribe by effectively forcing an exit from kinship.[62] This further induced a shift of identity away from the tribe toward the state. In this case, the state was not the usual political state but rather the social one of the Church.

The foregoing quotation also shows that the Church expanded the territory of its civilizing role as the referenced tribes were outside the Roman Empire at its demise. The Church's territory expanded after the fall of the Roman Empire and thereby increased the ultimate size of what was to become Western civilization.

Evolution of the feudal system. Before Roman conquest, tribes had their system for producing and distributing food. Upon conquest, and with the conqueror taking ownership of the land, some other arrangement had to be made to serve this function. Additionally, someone had to work the land to provide food for the people, and labor is not free. These two necessities led to what was to become the manorial system.

The landowner, who was usually a member of the hereditary elite, is dependent on his land to meet economic needs, and for this, labor is required. The laborers or peasants need access to land so they can produce for their needs. To meet the needs of each, the manorial economic system arose during the Roman Empire and became the organizing principle for its rural economy.

Manorialism vested legal and economic power in the landowner, who in turn received economic support from the manor's production. Production came from the peasants, whose labor was a legal, obligatory requirement, as, by law, they had to work for the landowner. In return for labor, the landowner granted the peasants rights of use to the land. Importantly, the law bound these peasants, or serfs, to the land. They could not leave the service of the landowner without his consent, making manorialism near-slavery. This was the start of serfdom. This system had variants and was more complex than stated. The main issue is that the lords had legal jurisdiction over their serfs. The state did not have jurisdiction,[63] making the serfs effectively noncitizens in their own country. Moreover, not only did the serfs have only minimal rights but their lords could change the rules. For our analysis, the peasants were legally bound to the land and its lord, and lacking mobility, they could not be part of any Differencing Engine.

Manorialism solved the food production problem. However, there was another issue affiliated with the Roman Empire, one associated with its fall and concerned with the state as a polity providing security. The Church does not provide security for this life, and as the power of the political state diminished, it left behind a security vacuum:

> Neither the State nor the family any longer provided adequate protection. The village community was barely strong enough to maintain order within its own boundaries; the urban community barely existed. Everywhere, the weak man felt the need to be sheltered by someone more powerful. The powerful man, in his turn, could not maintain his prestige or his fortune or even ensure his own safety except by securing for himself, by persuasion or coercion, the support of subordinates bound to his service.[64]

Before the state and empires like the Roman, responsibility for security fell to the kinship groups. After the Roman Empire fell, a requirement for security remained, but there was no entity to provide it; the kinship group was long gone. In this power vacuum, feudalism evolved from manorialism and, according to Fukuyama, was an alternative to kinship[65] that arose out of necessity.

Amending the manorial system solved the problem of security due to the absence of the state. The step from manorialism to feudalism involved adding mutual obligations and responsibilities to each party. Even though they are separate, feudalism was essentially manorialism with a security function.

Manorialism was for food production and distribution; feudalism was about security. Manorialism was about exchanging labor for rights to land, and feudalism added to this mutual obligations relating to security. It is a social arrangement that solves the political problem of security by modifying an existing economic system. We do not want to expend too much more effort here; even the experts cannot agree on a definition of feudalism. In the end, everyone gets food and security, and the serfs are still legally bound to the lord and his land.

In a tribal society, kinship is the tie that binds. Individuals are not free; they are bound to the group by blood. With the feudal system, the tie that binds is different. It is the contract along with mutualism, not blood, that ties men to one another. In both cases, tribal and feudal, bound individuals lack freedom, a requisite for the Differencing Engine, which has mobility as a requirement.

The Church as state. When the state of Rome fell, what should have been a vacuum did not materialize. There was a second parallel state waiting in the wings to take over—it was the Catholic Church, and it was ubiquitous in the old empire. It had as its principal concern security in the hereafter for its parishioners and was primarily a social institution and neither an economic nor a political one. However, with the absence of state, to maintain continuity and civil order, the Church assumed expanded functions. The following quotations from Pirenne attest to the Church's expanded role:

> When the disappearance of trade, in the ninth century, annihilated the last vestiges of city life and put an end to what still remained of a municipal population, the influences of the bishops, already so extensive, became unrivalled.[66]
>
> In addition, the bishop enjoyed very loosely defined police powers, under which he supervised markets, regulated the levying of tolls, took care of bridges and the ramparts. . . . In short, there was no longer any field in the administration of the town wherein, whether by law or prerogative, he did not interfere as the guardian of order, peace, and the common weal. . . . The populace was governed by its bishop and no longer asked to have even the least share in that government.[67]

We will shortly see that Western civilization is primarily a product of cities that arose during the Middle Ages. Since they came about due to the merchant class, we will refer to them as "commercial cities." Importantly, it was the Church, by assuming many functions of the political state, that enabled this process.

Transition to the genesis of civilization. We have arrived at an important point in our story, the time after the fall of the Roman Empire. This is the period when the Church is the only effective state. It is a time of deurbanization when cities shrink, even to disappearance. This is the start of the medieval period, when society is in transition, a time of serfs and feudal lords, and the time before reurbanization. Our Differencing Engine needs cities with specialization. It does not function with the immobile serfs.

The peasant serf might have been a parishioner of the broad state of the Church, but she was not a citizen of any political state. Being a citizen implies being subject to the rule of state, but, politically, serfs were the subjects of the lord of the manor and his laws. They were neither citizens nor free.

The foregoing is important. Civilization is dependent on citizens being responsive to the state. As long as a preponderance of a population is subject to an institution or entity other than the state, it cannot be civilized. Such was the case with serfdom, and this was the condition of the majority of the population.

Western civilization could not become civilized until serfdom's end. It was not until the twelfth century that serfs in Western Europe started to gain freedom to various degrees and differing times.[68] We have previously noted that in England, the end of serfdom started with the Peasants' Revolt of 1381. A metric for this change is that by the time of the French Revolution in 1789, the former serfs owned 50 percent of the land in France, double that owned by the nobles.[69]

Logic dictates two classes of people needed for civilization. It must encompass the vast majority. Can a state be civilized if 90 percent are slaves and the remainder freemen? I suggest not, and this is where the problem of serfdom comes into play. They are not citizens of the state, only slaves of their lords. The other class is not actually a need but rather a consequence of civilization's primary requirement: the city. Cities are hierarchal and consequently have a middle class or bourgeoisie. This new class comes with cities founded on commercialism during the Middle Ages and is the topic of the next section.

Western civilization comes with the new cities. As we have noted, when the Roman Empire fell, so, too, did its cities. With deurbanization, and as we saw illustrated earlier in figure 12.2, cities essentially ceased to exist. However, it seems as if, once started, the economic activity of trade and commerce never ends. Once this leverage into prosperity commences, it finds a way to keep going, and this was the case after the fall of Rome.

The nadir of collapse was reached sometime around the ninth or tenth century. This was followed in the towns and burghs with a positive inflection point caused by the commencement of an uninterrupted growth of merchant groups. This started in the tenth century, and by the twelfth century, they had grown so much that new parishes had to be created for them.[70]

The scene is set. There is the Church. Various states exist, but they are more concerned with outside forces than with internal ones. Besides, they are weak and the feudal lords compete with the weak state. The greatest percentage of the population consists of serfs, slaves of their lords. There is a new class, the middle class, starting to form. There are no or very few cities. There is no civilization. The time is the tenth century, and the place is Western Europe.

At this time, the main economic power resided with the lords and their manorial organizations. Having a feudal overlay, they were also a political power. Consequently, they competed with the political power of the state and its king. This tension between these two powers, as we will shortly see, aided the creation of a new institutional power, the city of commerce.

Commerce and trade are not agrarian activities. They require cities, and logic places them at ports and along trade routes. Logic is right; that is what happened. These cities needed physical security, and this they got with walls and resident knights. What the city did not have, or rather what its main denizens, the businessmen of the day, did not have, was legal security. Business needs the grease of certainty that comes with enforceable rules for commerce. This is something that comes with the modern state, but the state of the tenth century had other things about which to worry.

When institutions do not meet the needs of men, men must make their own. This they did in the making of the semi-autonomous city, a city having as a purpose commercial activity. This was not a top-down activity directed by the state or the Church; it was bottom up, directed by middle-class men. Importantly, these early businessmen made what best served their needs. After all, people do what is in their own best interest, and they created a commercial system fitting their needs. The system they designed and built was not for the State, Church, or feudal lords. Importantly, as Marx understood, it also was not for the peasants, who later were to become the proletariat. It was for the bourgeois themselves and their capitalist-based livelihood.

These new cities had a complex relationship with the existing powers, primarily for the reason that they represented a threat to existing power structures. The following quotation from Adam Smith illustrates the issues:

> The lords despised the burghers, whom they considered not only as of a different order, but as a parcel of emancipated slaves, almost of a different species from themselves. The wealth of the burghers never failed to provoke their envy and indignation, and they plundered them upon every occasion without mercy or remorse. The burghers naturally hated and feared the lords. The king hated and feared them too; but though perhaps he might despise, he had no reason either to hate or fear the burghers. Mutual interest, therefore, disposed them to support the king, and the king to support them against the lords. They were the enemies of his enemies, and it was his interest to render them as secure and independent of those enemies as he could. By granting them magistrates of their own, the privilege of making bye-laws for their own government, that of building walls for their own defence, and that of reducing all their inhabitants under a sort of military discipline, he gave them all the means of security and independency of the barons which it was in his power to bestow.[71]

The semi-autonomous cities thus created by the burghers represent a new institution, the municipality created to serve the commercial interests of the middle class. They exist for the sake of a new player, one that is not a hereditary lord or member of the ruling family.

We now have what we call in modern times the middle-class businessperson. Other names for this person are "burgher" and "member of the bourgeoisie." The new players, the nonhereditary middle class, make the new institution, the commercial city.

The making of this new class, the bourgeoisie, is discussed by Pirenne:

> Little by little the middle class stood out as a distinct and privileged group in the midst of the population of the country. From a simple social group given over to the carrying on of commerce and industry, it was transformed into a legal group, recognized as such by the princely power. And out of that legal status itself was to come, necessarily, the granting of an independent legal organization.[72]

I view this new class of the bourgeoisie, the men creating these commercial cities, as the founding fathers of America's founding fathers. City by city, they came together to form a city-state within the larger state, one that met their commercial and security needs. Importantly, these early founders, in making cities of their definition, established the precedent and process for creating bottom-up institutions, ones that can stand against other institutions like the state.

Just like the superorganism when it acted as dominant authority, the burghers through collective agreement decided on rules they needed. They filled in the void not covered by either the weak state or strong Church. They added to what was by forming their own institutions, ones they designed to meet their own needs and not those of others.

This could never have happened in China. Authoritarian governments would never permit such foolishness; they keep absolute control of absolutely everything. Moreover, this probably could not have happened unless Rome fell along with there being a nonpolitical state, the Church, available to pick up the slack. Requiring empire's fall plus the existence of a surrogate is contingency at its best.

The powers are now the Church, the State, and the feudal system of lords and serfs. The newly ascendant commercial city with its burghers is the new power. It is a power of the people, specifically the commercial middle class.

The Church might have played midwife, but it was not the mother of Western civilization. Civilization comes from cities. In the case of Europe, it was uniquely a bottom-up proposition. The civilizing process might have started with the Roman Empire and then continued with the Church, but, in the end, civilization came from the city. The city in this case was not a top-down creation of the state. It was bottom up from the people; it was from the burghers. It was the bourgeoisie. It was the middle class. Using the language of modern leftist political discourse, the foundation of Western civilization was privileged white males who also happened to be capitalists, greedy or not.

Consequence of civilized cities. Being semi-autonomous, these commercial cities were able to confer social status and political rights on their residents.[73] This power was highly consequential, as it included the power of granting freedom; it could free men from their bondage of serfdom. The city could break the bond between serf and lord, making the serf a freeman. "It made the cities gateways to freedom, holes in the tissue of bondage that covered the countryside. *Stadtluft macht frei* ran the medieval dictum—city air makes one free."[74]

In the foregoing, we see that cities played a role in ending serfdom, obviously placing them in conflict with the feudal lords. Additionally, they competed with the territorial lords for human resources. The feudal lords and the emergent commercial cities were natural antagonists.

Commenting on the status of the individual in the new city, including former serfs, Pirenne states,

> Birth meant little. Whatever might be the mark with which it stigmatized the infant in the cradle, it vanished in the atmosphere of the city. This freedom, which at the beginning only merchants had enjoyed *de facto*, was now the common right of all burghers *de jure*. . . . Freedom, in the Middle Ages, was an attribute as inseparable from the rank of citizen of a city as it is in our day of that of a citizen of a State.[75]

Freedom is sometimes an abstraction. In the new commercial city, it was not. The escaped peasant of the Middle Ages could achieve freedom by reaching the ramparts of the city. Once within this new institution, all men are free and, by this nature, draw other men wishing to be so.

This is fertile ground for the Differencing Engine. All the requisite elements are in place. However, as Fukuyama explained, there is one more feature, the roots of democracy: "contemporary conventional wisdom has it that democracy will not emerge without the existence of a strong middle class, that is, a group of people who own some property and are neither elites nor the rural poor."[76]

We see here in the tenth century and the new commercial city with its bourgeois the foundations of a Western civilization to be, a civilization having freedom, capitalism, and the ingredients for democracy. If Rome had not fallen, Western civilization probably would have still risen. After all, civilization appears to be on humankind's natural trajectory. However, without the contingency of Rome's fall and the existence of the Church, what would be its nature?

East versus West. The development of civilization in the East by China's Qin Dynasty was a top-down process. It used coercive methods grounded in Legalism to achieve its unifying ends. Recall that Legalism embraces the concept of collective guilt, where the entire family pays for the guilt of one family member. This is a powerful tool with which to break kinship ties.

The formation of Western civilization was different. It was not the consequence of state coercion. Many factors we have discussed were in play, but the critical one is that it came with the development of the commercial city. It developed willingly and out of necessity for commercial enterprise; importantly, it was a bottom-up creation for the needs of the commercial collective.

Protestants. We are now going to split the Protestants from the Catholics. This adds to our analysis some aspects of culture important for wealth creation: the Protestant traits of hard work, literacy, and personal responsibility.

Figure 12.3, with one exception, is the same as figure 4.1, "Wealth of Nations," we have seen before. In this case, we replaced "Western" with "Protestant" and "Catholic," showing a GDP per capita $47,513 for the Protestants and $33,101 for the Catholics. We are about the causes of differences in economic performance.

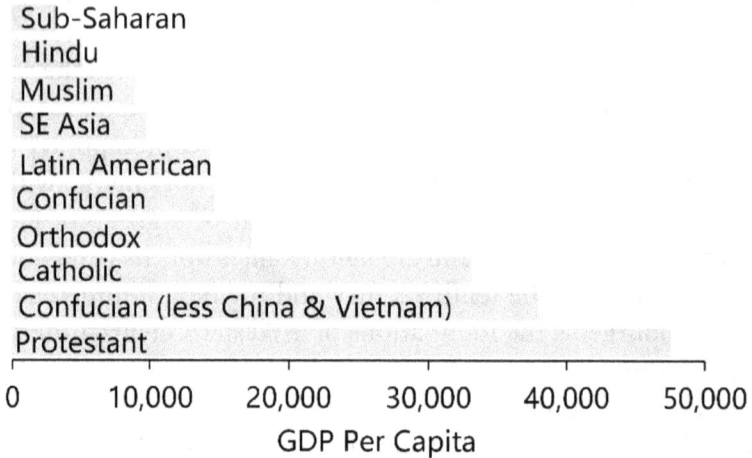

Sub-Saharan
Hindu
Muslim
SE Asia
Latin American
Confucian
Orthodox
Catholic
Confucian (less China & Vietnam)
Protestant

0   10,000   20,000   30,000   40,000   50,000

GDP Per Capita

FIGURE 12.3   Wealth of Nations, with Protestant and Catholic

Consequently, with the Protestants making 44 percent more than the Catholics, there is a story that we should understand.

In our discussion of Western Christianity, we found that the West became economically powerful owing to its cities being bottom-up creations designed for the capitalist mind-set and optimized for the Differencing Engine. They were designed to make money for the individual businessperson and not for the ends of an absolutist state, such as in China.

Changes of the Reformation. We have discussed that the Roman Catholic Church of medieval times was a state that provided security for the next life, not this one. By self-proclamation, it held the keys to the Pearly Gates, making it the ultimate intermediary between the individual and God. Following the dictates of the village priest guaranteed salvation, not following, damnation. It was a simple system, one in which the Church held all the keys to salvation.

In 1517, the publication of Martin Luther's treatise, *The Ninety-Five Theses on the Power and Efficacy of Indulgences,* challenged the power of the Catholic Church. It questioned the idea that salvation could be obtained with the purchase of indulgences, a scheme where the individual could pay money to the Church for absolution of sins. Luther's central argument was that salvation was individual, that is, between man and God. This became the central tenet of what was to become the Protestant faiths. Security for the next life became an issue of personal responsibility, reducing the importance of the Church. It no longer owned the keys.

We get with the Reformation a force for personal responsibility, a positive trait for economic success. Before, you dealt with the local priest and he directed your activities toward soul-saving ends. Now, there is no intermediary priest, and you have all the responsibility. You, the individual, have to decide.

This newly acquired responsibility for making decisions concerning salvation required individuals to know the Bible. This they needed to arrive at correct decisions as related to God's will. Before, this interpretative function was the sole province of the Church; after, it became that of the individual. Consequently, a second feature of the new paradigm was a requirement to read. People dependent on self for salvation are not going to depend on the third party Church in the form of the village priest for interpretation of God's will.

Protestants learned to read, Catholics not so much. Being in charge of your own salvation meant dealing with God one on one. For that, you had to be able to read the Bible. Protestants learned and in the process became literate, and literacy has a positive role in affairs economic.

A third consequential trait of the new Protestantism was hard work. It was not simply working hard to get the job done sooner than later; it was hard work for work's sake. There is a Protestant work ethic where hard work is its own reward, which, like the first two, adds to the economic success fostered by this alternative to the former monopoly of the Catholic Church. The work ethic became a feature of Protestantism. Work of itself was Godly; that was reason itself. This generates the "live to work" ethos so admired by Landes.

Power to the people. With their serfs, the lords of the rural realm retained their power over the old agrarian economy. The cities bring a new economy, one based on commerce, and it is here where the freeman burghers ruled. They had economic power, but that is not all. They were in charge of life at the municipal level, and this gave them political power. Finally, the city was its own social center, and there too the people ruled. In this new commercial city, power, political, economic, and social, was with the people. There should be no wonder that the hereditary lords detested these low-bred usurpers.

We see that before the Reformation, the burghers controlled three legs of their lives, economic, political, and social. There is a fourth leg, salvation for the next life, and the Church owned this. The burghers remained subject to that institution and had no say in it. All that changed with the Reformation, and people were now in charge of their own salvation. After the Reformation, people had power over all institutions. This is powerful. The people now have full control of their destinies, which is the proactive environment where change will happen.

Why are the Protestants so much more affluent than the Catholics? The preceding ideas point us toward where to look for the answer. Holding the ethic of hard work and personal responsibility along with being literate are traits that would help any individual obtain economic success.

Protestantism and its associated traits is not solely a property of individuals; these traits are the property of a culture. This is a difference with a distinction. Individuals can hold these properties and, by them, excel in economic endeavors. However, individuals are ephemeral beings with no persistence. Culture, a property of populations, is different. When Sowell writes of Germans and the importance of education, he states, "This high priority of education was more than a policy of government, it was a cultural value of a people."[77] Culture persists. The high economic performance of Protestants is only proximately due to its individuals. Ultimately, it is its entire culture along with culture's persistence.

The hubristic West. The West's hubristic view is that it is the best. It has earned that title, but reality changes. If we are to believe this book, then based on IQ scores, the people of the East are superior to those of the West. The East will soon be the best.

Obviously, there is more to this life than IQ scores, and "best" has several flavors. However, IQ is as good a predictor of future outcomes as any, and IQ scores predict that, in time, China and its Confucian cohorts will rule the economic world. The West better prepare. It best address this reality and figure out how it is going to work with the East, and it better do that before the East is top economic dog. It is better to control your own future than have it controlled for you.

Orthodox Christians

Introduction. The Orthodox nations have Orthodox Christianity as their religion. There are ten nations on our list, including the non-Islamic nations of the Balkans. However, the East Slavic nations of just three, the Russian Federation, Ukraine, and Belarus, account for 78 percent of the total population. Given its dominance, we will focus our attention on Russia and its history.

Early history. We started our discussion of civilizing Western Europe with the invasion of Great Britain by the Roman Empire in 43 CE. From that point, we have to skip more than eight hundred years just to see some Russian history, let alone its later conquest by the Mongols. Just as the West was a late starter on the path of civilization when compared to the Confucian East, Russia was later yet.

We start with the Kievan Rus', a federation of Slavic tribes that started in 882 in the land between the Baltic and Black Seas. Timewise, this corresponds with the nadir of the Dark Ages in Western Europe, just before starting its ascendency. The Kievan Rus' came on the scene well after the collapse of the Roman Empire, and although adjacent to it, they were never part of the Byzantine Empire.

An underlying reason for the Kievan Rus' was their location on three major riverine trade arteries, the main one being the Volga River. They were the gateway for trade between Northern Europe and the Byzantine Empire and beyond, where beyond reached to Central Asia.[78]

It is interesting to note that what is to be Russia starts with a commercial economy based on trade and freemen and later reaches an apex with an agricultural economy founded on the backs of serfs. The opposite sequence occurred in Western Europe. There a serf-based agricultural economy came first, followed by the development of a commercial economy that started with trade.

Trade informs culture, and in this case, Byzantium influenced the Kievan Rus' and the Russian Empire to be by sourcing Christianity to the Slavs. Orthodox Christianity became their religion, and it was from these roots, first established in the ninth century, that Russian culture, art, and literature evolved.

Western Europe was conquered and colonized by the civilized Romans. Even though Rome collapsed and Western Europe went into the Dark Ages, residuals of a grand civilization remained. The Eastern Europeans were not so lucky. In the late thirteenth century, they got for their conquerors the barbaric Mongols, a tribal people exhibiting the antithesis of everything civilized. This Golden Horde of Genghis Khan was like a gang of street thugs from the neighboring city taking over and demanding tribute for peace. They were barbaric to an extreme, killing all in a single town just to make a political point. However, during their occupation, the locals were allowed to run their own political operation—but pay tribute to their conquerors they must.

The Mongols were not only physically destructive; they also destroyed Kiev along with murdering most of its inhabitants. They killed progress, and in this, they had a huge negative influence on the civilizing process. They cut off communication and trade between Russia and their cultural mentors, Byzantium, along with lands farther to the east. They also hindered communication with Europe, consequently blocking any influence of either the Reformation or Renaissance. With the destruction of Kiev, the locals had to start over, and by the time the Mongols left, 250 years after conquest, the center of political gravity had shifted to the Grand Duchy of Moscow.

Serfs and serfdom are important topics to our narrative. In Western Europe, this institution commences with the Roman Conquest and, by the twelfth century, was in decline. In Russia, it is does not even get started until the thirteenth century, with the start of the Mongol domination in 1240.

When we arrive at the mid-fifteenth century, the Grand Duchy of Moscow has been the center of power since the late thirteenth century, and Ivan the Great is Tsar. It is time for the locals to get on with creating the Russian Empire and the civilization of Orthodox Christianity. They do. Ivan the Great finally rids the land of the Mongols and, acting as a centralizing monarchy, expands the territory of the state threefold.

In the late fifteenth century, when Ivan the Great gathered under a single authority various bits and pieces and thus created a unified Russia, he is like Qin Shi Huang the First Emperor of Qin and a unified China. The span of time between China's unification and the formation of Russia is approximately seventeen hundred years. We can use the time mark of unification as the starting time of the civilizing process. Whatever changes are effected on people and their nature by being civilized or not, seventeen hundred years is ample time for people to become different.

Even with ridding the land of Mongols and unifying Russia, Ivan the Great had his negatives. "Freedom was stamped out within the Muscovite lands. By his bigoted anti-Catholicism, Ivan brought down the curtain between Muscovy and the west. For the sake of territorial aggrandizement he deprived his country of the fruits of Western learning and civilization."[79]

Western civilization developed under a system of laws, constraining the power of the state that, even by medieval times, was well developed.[80] This is in contrast to China where, as exampled by their emperors' exercise of tyrannical powers, their civilization rose at the opposite end of the legal spectrum. According to Fukuyama, "they engaged in wholesale land reform, arbitrarily executed the administrators serving them, deported entire populations, and engaged in mad purges of aristocratic rivals."[81]

Russia was just like China. In the wide spectrum of state behavior, it too was at the extreme, authoritarian end. Furthermore, it was the only European state exhibiting such despotic, unconstrained tyranny.[82]

There is another contrast. Before the fifteenth century, when compared to its West European neighbor, East Europe was an underpopulated frontier. In this sense, it was like early America with its open, unused land presenting great opportunity. These sparsely populated lands resulted in less state control and, because they offered individuals more freedom, induced colonists from the West as "colonists from the Western Europe and Eurasia could live under their own laws."[83] Back then, the Russia to be was the land of the free, a condition that was about to change with the genesis of the Russian Empire.

Starting in the fifteenth century, this *laissez-faire* frontier attitude changed. Russian serfdom takes a nasty turn, with the state issuing new rules restricting the freedoms of the peasantry. Specifically, the Russian Muscovite state placed severe restrictions on their movements, and they were required to remain with their land. This effectively shackled them, just like slaves elsewhere. The consequence of violating these restrictions was big fines for both the peasant and for any others providing assistance. Adding insult to injury and concurrent with the further constraints of the serfs, there was an increase in the rights of their lords.[84] Russia is now feudal and its serfs slaves.

The European West was different. Although it was also feudal, it had free cities, which made maintaining serfdom a difficult proposition. So, while freedom was increasing for peasants in the West, those in the Russian East were being placed in increasingly restrictive yokes,[85] ones that were to remain well into the nineteenth century.

A Tsar is the head of the Muscovite state and the landowners are the nobility. The vast majority of the population is serfs, slaves owned by the landowning nobility. There is essentially no middle class and little commerce. There is the strong authoritarian state, a noble class aligned with the state, and an Orthodox Church symbiotically dependent on the state. This is the Russian Empire's starting point.

The state of the Tsars. We stated earlier that all rich nations have civilization in their history but that not all civilized nations became wealthy. Russia is an example of the latter; it is civilized and yet is relatively poor. The reason why Russia is poor is the story for this section, which starts with the founding of the Russian Empire.

In the prior section, we discussed that what was to become the Russian Empire started in the fifteenth century with Tsar Ivan the Great. It finally ends in the early twentieth century with the revolution heralding in the Communist state and the Soviet Union. Even though empire ended, the authoritarian state kept its ways with the new Communist state, and the peasants simply exchanged one nasty boss for another.

One of the issues we have discussed about becoming civilized is the change of identity to the new state from the old tribe. It was under this paradigm that we discussed the Catholic Church attacking certain aspects of kinship rules, resulting in a shift in inheritances favorable to the Church. The consequence of altering inheritance practices was to weaken the tribe.

We do not know the intent of the Church's actions, only its consequences; it got richer and the tribes weaker. What is important for our narrative is that a similar situation did not occur with the Orthodox Church.[86] On its lands, kinship retained its strength for longer, and the change of identity from group to state was delayed. Unlike the Catholic Church in the West, the Orthodox Church probably had little role in the change of identity. In Russia, this change was probably due to the nature of the authoritarian state.

We have defined the civilized as comprising societies where individuals identify with the state and not the group or tribe. Strong authoritative states, where the nobility or lords own the peasantry, break tribal bonds and force individuals to identify with their owners. The Russian state under the Tsars was such a state, defining Tsarist Russia as civilized.

According to our model, for a nation to become rich, it first needs be civilized. However, set deeper within this requirement is the need to have our Differencing Engine running, making cities and their commercial activities requisite.

As pointed out in the following quotation, cities are more than the place country folk move to from the farm. Importantly, they are the starting point or kernel of the process where an agrarian economy changes to a commercial one, one where the Differencing Engine runs:

> Cities and the bourgeoisie, then, do not simply come into being as a result of economic growth and technological change, as Marx believed. They are initially weak and vulnerable, and unless they are granted political protection, they will be subordinated to the pow-

erful territorial lords. This is exactly what happened in Poland, Hungary, Russia, and other lands east of the Elbe, where a different configuration of political power either made monarchs weak or induced them to side with one or another stratum of the aristocracy *against* the interests of the townsmen. For this reason, there was never a strong, independent bourgeoisie in Eastern Europe. A technologically advanced capitalist market was not introduced by townsmen but by progressive landowners, or by the state itself, and therefore failed to flourish to the same extent.[87]

In our discussion of the Protestant religion, we noted that cities founded on commercial enterprise started to rise in the West during the tenth century. These were the starting point of a capitalist class and the modern economy of the West. In the following, Fukuyama explains the underlying reason:

> But there was a precondition for the rise of a capitalist class in the
> first place—the mutual hatred of the townsmen and the king for the
> great lords. Where this condition did not prevail, as in many parts of
> Eastern Europe, no such class emerged.[88]

The commercial city, to gather power, especially for establishing rules relating to commercial enterprise, needed freedom to do so, and this can only arise if opposing powers are kept at bay. In the West, the weak state sided with the nascent burgher class against the strong forces of the territorial lords. This enabled the creation of self-sufficient municipalities as political institutions with their own legal jurisdiction and with political power.

The state of Tsarist Russia was different; it was authoritative and strong in its own right. The Orthodox Church, having been cut off from its Byzantine origins during the Mongol occupation, found protection in the state, and in return, the Church gave the state legitimacy.[89] Add to this symbiotic couplet a relatively weak aristocracy that was already coaligned with the state and we have a very powerful state.[90] It is one where all three institutions speak with a single voice.

Because those having power do not willingly give it up, any new group seeking a share of power, such as an embryonic commercial or bourgeoisie class, could expect resistance. The consequence was that in Russia, any capitalist class that developed could not acquire power, and a modern economic system remained unknown.[91]

A quotation from Sowell provides further insight into this issue:

> The social milieu in czarist Russia was one in which the nobility looked down with contempt on business and businessmen. With a vast and illiterate peasantry, a class of nobles who contribute little to commercial and industrial development, and an intelligentsia hostile to capitalism, it can hardly be surprising that Russia had to rely heavily on foreigners to modernize its economy.[92]

In the West, businesspersons developed power and reached parity with other institutions. In Russia, they were looked down on, even to the point of contempt, and any group having such social standing cannot gain respect, let alone power. In addition to the specific problem of not being able to form a commercial class with power, Russia, like its Slavic cohorts in Eastern Europe, was a backward country when compared to its European neighbors to the west.

One of the great figures of Russian history is Peter the Great, the Tsar of the Russian Empire from 1682 to 1725. From the perspective of our narrative, his main impact was a result of recognizing Russian backwardness and taking significant action to correct this deficiency. He went as far as to tour Europe in the disguise of a worker.[93] He both sent Russians to the West to learn and brought from there human capital to teach. These imports probably played an important role in the commercialization of Russia's major cities.

Even with an enlightened Tsar, Russia had a long way to go to develop a commercial class:

> Between the nobility and the peasantry there was little middle class. Businessmen were regarded not so much as assets to the country but as prey for government—both the official treasury and individual corrupt officials.[94]

Peter's effort at modernizing Russia was what would be expected of a Tsar used to having his way, authoritarian and top down. It was of the command variety—the state knows best, so do it! It was not an organic effort of a natural, bottom-up approach, the type used by the burghers of Western Europe in creating their new municipalities.

Even with Peter the Great's effort to bring modernity to Russian enterprise in the early eighteenth century, modern industrial development did not commence until the late nineteenth century.[95]

Russia in transition. In 1861, just before the arrival of the twentieth century, Emperor Alexander II of Russia abolished serfdom. All it takes are a few numbers to provide a sense of the immensity of serfdom in Russia:

> The upper reaches of the Russian nobility were staggeringly rich: Count N. P. Sheremetov owned 185,610 serfs, while his son, Count D. N. Sheremetov, managed to increase that number to more than 300,000.[96]

For purposes of perspective, note that in 2013, Microsoft had fewer than one hundred thousand employees.

Entering the twentieth century, the slavery that was serfdom in Russia had ended, but there was little change in the nature of peasant lives. They remained tied to their sole source of sustenance, the farm, and Russia was still primarily a land of peasant farmers.

By comparing the Russian class social structure to that of England for the period at the end of the nineteenth century, we can learn a lot about Russia in this transition period. This will tell us much of what we need to know.

Figure 12.4, even though it makes an apple-to-orange comparison, will be our storyteller. One source of data, the 1881 census for England, has five classes based on the occupations of a fully industrial society. The second source of data, the 1897 census for Russia, also has five classes, but they are ones based on a mixed feudal and industrial society. Even with this minor dissimilarity in classification, the data are informative.

The fact that Russia had very few engaged in commercial activity, the commercial and worker class, speaks volumes about its economy; it was not much. However, the main part of the story is the difference in the unskilled/peasant class. While Russia had 82 percent of its population in this class, the corresponding value for England was only 10 percent. If Russia had been more like England, these numbers would be similar and Russia would have had considerably more in the commercial and working class than it

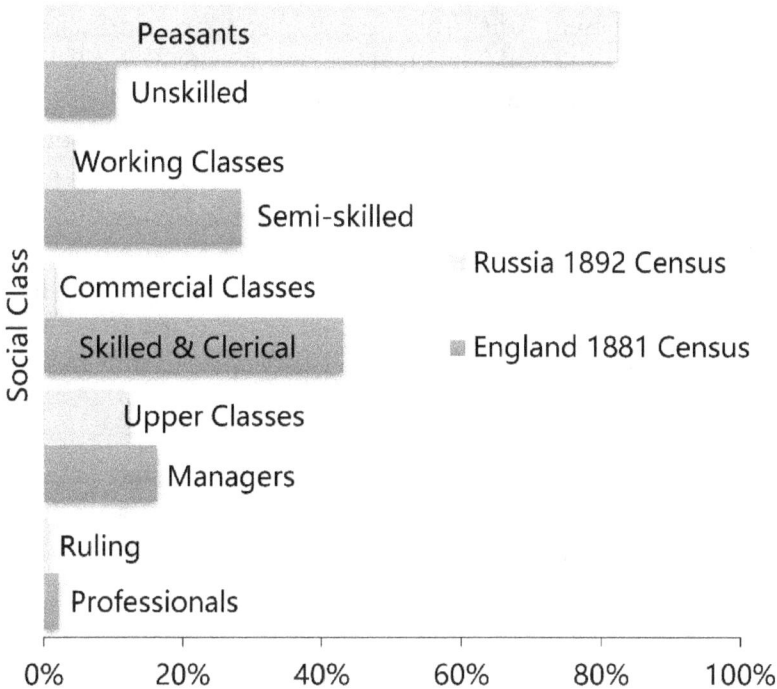

Peasants
Unskilled
Working Classes
Semi-skilled
Commercial Classes
Skilled & Clerical
Upper Classes
Managers
Ruling
Professionals

Russia 1892 Census

England 1881 Census

Social Class

0%   20%   40%   60%   80%   100%

FIGURE 12.4   Social Structure of England versus Russia

did. This disparity is indicative of Russia's weak commercial and industrial status. The census shows that as late as 1897, almost everyone in Russia, or at least 82 percent of its population, was on the farm, and workers on the farm do not make tractors or smelt iron.

We can consider figure 12.4 as a picture of Russia's six-cylinder Differencing Engine running on only one. Consider that its civilizing process started with the first Tsars and the Grand Duchy of Moscow in the late thirteenth century. Moreover, six hundred years later, in 1892, it is only at figure 12.4 where 82 percent of the population remains engaged in what is essentially primitive subsistence agriculture. This leaves only 18 percent of the Russian population participating in its barely running Differencing Engine. Additionally, given what we have discussed about the social standing of businesspersons, even that part of the engine needs help.

310   INEQUALITY

Is this lack of economic success due to people or lack of comparable development time? The West did get a kick-start with the Roman Empire, which left residuals of civilization after it fell. It is difficult to pin down the start of Western civilization, but sometime after Rome's fall and during the Dark Ages is a reasonable guess, and the fifth century is a reasonable candidate. This leaves Russia starting its civilizing process eight hundred years later than the West, a considerable lag time. However, if we assume the civilizing force of Rome to be the West's actual starting point, that increases the time differential to thirteen hundred years, a significant span of time. Whichever we choose, the difference in time in being civilized must be a major factor.

Another factor is the Russian history of authoritarian rule. The state in the West was never as strong as the Russian state. It had to compromise, where all the players got something. This allowed the will of the people to gather power and become an institution in its own right, something that could never happen in authoritarian Russia.

A third factor is the Catholic Church. Mainly, it provided continuity for what Rome started. In contrast, Russia got a kick in the pants from its conquerors, the Mongols, not a kick-start.

Russian peasants. Other than being rural, who were Russia's innumerable peasants, these recently freed serfs? From figure 12.4, we see that about 82 percent of the population in 1892 was peasants and that "as late as 1897, only 21 percent of the population of the Russian Empire was literate."[97] Therefore it is reasonable to assume that there were few literate peasants.

While the condition of literacy has importance, behavioral motivation has more. We can get insight into the motivation of the Russian peasantry from the works of the Russian economist Alexander V. Chayanov (1888–1937), who "claimed that consumption, or the family's subsistence, was the motive determining the peasant's activity."[98] It was from this thesis that Chayanov developed an economic theory of peasant farm labor, the principle of labor–consumer balance between meeting family needs and the drudgery of labor.[99]

Chayanov's peasant farmer did not engage in capitalist activities. Production was for family, with the family providing the labor; they did not use hired labor. They did not take their produce to market; they consumed it. Labor was only sufficient to meet needs and not more. Leisure had value.

This Russian peasant farmer was similar to the Indonesian peasant discussed earlier. The main difference was that, in Indonesia, we discussed the individual as part of a village, whereas in Russia, the extended family replaces the village. Otherwise, they are alike.

We have in the Russian peasant farmer something new to our narrative. The economy of this "modern" peasant farmer was not tied up in social relations, as were subsistence farmers of the superorganisms. Recall that labor for this latter group was for the good of the group and not the individual; it was all about social relations. In the case of the Russian peasants, the farmers were in it for themselves and their families.

The motives for labor in the superorganism were to satisfy social requirements, and owing to the nature of the social organization, this labor satisfied the needs of the group. Recall that working for living and not profit explains and describes the Indonesian villagers we have discussed earlier. They likewise only worked to meet their needs, and once these needs were fulfilled, they stopped. They had other things they would rather do. The motives for the Russian peasant were to satisfy needs directly. The similarity between the two classes, the Indonesians and the Russians, is that neither had profit as a motive. Once needs were met, work ended.

When profits are the motive, there are no limits, but with satisfied needs as the objective, the limit is satiation. Consequently, the Russian peasant farmer does not add anything to the Russian economy. This is just as the Indonesian villages we discussed, who also do nothing for their economy.

Were the peasants of late-twentieth-century Russia organized as superorganisms, as proposed for the Indonesian villagers? The answer is probably no; such societies do not survive in strong, authoritative states, such as those represented by the Tsars. The Russian peasants are different in their social sense from the Indonesian villagers.

This question may or may not be important, but the fact that the Russian peasants behaved similarly to pre-state, superorganismic groups is. From the perspective of human economic development, this places them, along with similar other subsistence farming peasants, on the wrong side of economic development, as the Differencing Engine requires an aspiration greater than just satisfying needs. In this aspect, the Russian peasants were just like other primitive, pre-state, illiterate societies.

It is interesting to note that as late as 1990, 50 percent or more of Russian agriculture remained as subsistence farming; it had not changed over to commercial.[100] Stalin and the Communists may have yanked the peasants onto collectives and forced major changes, but a reduction of only 50 percent from subsistence farming suggests that the Communist reform was ineffective in changing the old ways.

The foregoing begs the critical question as to what is the best method to convince subsistence farmers to change. How do you get people to want stuff to the point that they will willingly work for it? This, I suggest, is the critical question for anywhere subsistence farming remains a way of life, one we will further discuss. However, in answering this question, we will have to be careful in separating the two different classes, one represented by the Indonesian villagers and the other by the Russian peasant.

The following quotation suggests that subsistence agriculture, as practiced in Russia, is a survival strategy. Contrary to the superorganism, its basis is independent of social structure:

> Russia's rural population is faced with shortage of alternative employment opportunities, extremely limited mobility of labor, and total lack of political lobbying power. These factors have produced a specific Russian model of agricultural labor markets in which unemployment is held in check by reducing the effective working time and lowering the wages. As a result, there has been a massive shift of agriculture labor from wage employment in corporate farms to subsistence and semi-commercial farming on household plots. Russian agriculture is thus characterized by large-scale underutilization of labor, primarily in subsistence household farming, which only deepens rural poverty by creating an illusion of employment, but without sufficient income.[101]

The preceding quotation gives possible insight into why the Russian peasant stops work once meeting needs. The economic milieu fails to offer opportunity for personal gain through profit from labor. A "shortage of alternative employment opportunities" combined with "extremely limited mobility" translates to being stuck with no available path either out or up. All the peasants can do is survive, and that is exactly what they do with subsistence farming.

Support for this view is in the following quotation:

> 40 percent of all Russian families engage in private subsistence farming. In the majority of cases, income from this source remains low. Private subsistence farming is a survival model rather than a tool for entrepreneurship and vertical mobility.[102]

The foregoing discussion does not make Chayanov's labor–consumer balance principle wrong or not useful. Its fault is that it does not contain an analysis of motive. However, as a descriptor of both the Russian peasant and the Indonesian villager, it does just fine. Empirically founded models such as Chayanov's do not answer the all-important "why" question and therefore lack depth.

The main point in the preceding discussion is that the Differencing Engine requires mobility, both up and down. If the economic system lacks an "up button" on its elevator, nothing else matters. The engine does not run in with an elevator, and that is where the Russian peasant is stuck.

There is another issue with the Russian peasant. Modern economies have as a major driver the consumer, and by definition, the subsistence peasants of Russia were not consumers. Consequently, with more than 80 percent of the nation's inhabitants being non-consumers, the economic performance of Russia in 1897 was far less than it should have been.

There is also a story in the 18 percent who are not peasants. We will assume that the 18 percent not on the farm are engaged with the modern economy. This suggests that whatever economic status Russia has obtained comes from this relative small portion of the urban population, as the city is the primary locale of any Differencing Engine. Recall that to change a population requires most to be part of the process, and a paltry 18 percent is not enough for the "population." Still, some changes occur, and the urban portion of the Russian population probably changed accordingly.

Changing direction with food security. We noted earlier that 40 percent of Russian families engage in subsistence farming. This is a huge number, possibly having significance to our story. It suggests that the collective memory of a prior collectivized Russia recalls an uncertainty of food security. They recall a period of insufficient sustenance. Planting turnips in the backyard insures against the incompetency of the state in ensuring availability of sustenance. It is a good example of the rational actor at work.

I have no idea how the noncommercial farming activity of the 40 percent translates into national GDP, but its influence must be substantial and significant. If Russia is to make significant economic progress, this is an area ripe for the picking. If people had faith in the state, far fewer would feel compelled to such activity and would redirect more of their energy toward activities resulting in an increased GDP. Memories are long and, once lost, faith is slow to return.

Top-down, authoritarian states do not lead to confidence. Too few can screw up too badly to gain trust for one's existence. This is additionally true in an oligarchy, where much of the economy is in just a few hands. Until Russia changes to a true liberal democracy and rids itself of its oligarchic ways, its people will remain wary, and rightfully so. They will act to protect their long-term survival, even if that action is not in their best short-term economic interest. When people are in charge of their own destiny, they need less protection against the incompetence of the autocratic state.

The foregoing is speculative. However, whatever the cause of so many engaging in subsistence farming, it has a large effect on the economy. The Russian state needs to understand why so many are engaging in non-GDP-producing economic activity and take appropriate action. Russian leaders insist that Russia is the equal of any on the world stage, but the 40 percent puts a lie to that insistence. Russia remains as it has always been: backward. That is too bad; its people deserve better.

Latin Americans

Introduction. We have now covered seven of our eight groups, and in doing so, we have covered most of the inhabited continents. All that remain are the southern half of Africa, with the sub-Saharan Africans, and South America plus the southern half of North America, which encompasses Latin America, the topic for this section.

Latin America, the most unique of all our eight groups, has some pure ethnicities, but in the main, it is a mongrel resulting from multiple crosses of many ethnicities and their cultures. It is not Inca, Aztec, or Mayan, its three civilizations, and it is not any of the multitude of its noncivilized, indigenous tribes. It is not Spanish or Portuguese, the two nations that colonized it, and not sub-Saharan African, a major constituency of its largest nation, Brazil. Nor is it Italian, a significant ethnic group of Argentina. It is those entire, but most importantly, it is the multitude of begats resulting from a seemingly infinite array of interesting, interethnic couplings.

Our model has served us well to this point. Using it to examine a people over an extended stretch of time has yielded insight into their current economic condition, either rich or poor. However, its utility diminishes with the Latin Americans because they are not a people of unified, continuous history. Their story starts as the others' do, but as we will shortly see, it disintegrates into a hodge-podge of discontinuities resulting from the European conquest.

Geography of the New World. A main difference between Eurasia and the Americas was the lack of domesticable animals in the latter. It had nothing comparable to horses, cows, and sheep. They are not so important for food, as calories and protein are easily encountered throughout the region. The central issue is animals domesticable for work and transportation. Another, often under-considered point relating to domesticable animals is that they are a source of disease, and disease is needed to get protection from disease.

The path to civilization runs through the farms of the Neolithic Transition, and this means working the land. Without oxen or similar beasts, all land preparation, including tilling, must be done by manpower. A farmer with oxen can produce much more than one without. Getting to cities and civilization requires food production in excess of the producer's needs, and those with work animals will get to that developmental point sooner.

The civilizing process usually entails war and its associated consolidation of groups. Large, powerful groups conquer and absorb smaller ones, and the conqueror's job is easier when there are animals available for transportation. Walking to war is not efficient and probably arduous. Additionally, having to walk constrains the size of the area available for conquering. The Khan family could not have traveled from Ulan Bator to Moscow and beyond on foot; the horse enabled their conquering ways.

War is not the only means of uniting disparate groups. Trade, by coupling disparate groups, has a civilizing role. Additionally, it fosters cultural exchange and spreads good ideas, including good food. However, trade requires communication, meaning that "Italians could acquire spaghetti from China but the Iroquois could acquire nothing from the Aztecs, or even be aware of their existence."[103] Furthermore, trade requires transporting goods, a task made far easier with animals.

No one enjoys getting sick, but there is a positive side to becoming ill. Humans get a variety of diseases from the animals they live and work with and, in time, build immunity. Immunity provides protection from disease, but unfortunately, you have to get sick to acquire it.

So, although getting sick is not good, acquiring immunity is. This was the lesson learned when disease-ridden Europeans entered the relatively sterile and disease-free lands of the Americas. Almost all the native hosts died, whereas the visiting Europeans suffered almost no ill effects from their contact with the new environment.

The physical geography of the Americas has two features detrimental to the civilizing process. It has a north–south orientation, the opposite to Eurasia's east–west. Recall that plants grow in bands circumscribed by common latitude. A plant domesticated in China will do quite well in France, but one from temperate Nebraska will never bear fruit, literally, in tropical Venezuela.

The Americas also have their road-blocking barriers, though perhaps not as dominant as those of Eurasia. There are the Rocky Mountains, the Sonoran Desert of the southwestern United States and northwestern Mexico, the swamp and jungle of the Darien Gap of Panama, and the Andes Mountains, a seven-thousand-kilometer mountain chain stretching from Chile to Venezuela. What was to become Mexico had two civilizations, Maya and Aztec, and north of that future border with the Yanquis there were none. There was never a Silk Road for transporting transportable culture.

Preconquest civilizations. Who were the people conquered by the Europeans at the end of the fifteenth century? Because almost all were to die from imported diseases in the century following first contact, is the question important? The answer is yes, for a couple of reasons.

The Americas were home to three civilizations, the Maya and Aztec of Mexico and the Incas of the western edge of South America. The fact that civilizations with properties similar to those in Eurasia arose in lands completely isolated until first contact makes a very important statement about human nature. It suggests that civilization is a natural waypoint of human development. The only requirement for its creation in any given population appears simply to be the appropriate environment.

Becoming civilized or not has nothing to do with the people themselves; all populations have the genetic code to become civilized, provided the applicable forces for natural selection are in place. Whatever code it is, it is in the population of all peoples, and in this, we are the same. That is not to say that resultant civilizations are equal, only that similar structures arise from a population of humans given appropriate environments.

This concept takes the proper sense of Jared Diamond's environmental determinism. If we all had the same environment, we would be equal. However, we did not and are not. What is the same is our universal potential to become civilized; that was and remains equal.

Even though the native populations of the Americas almost vanished because of diseases brought with the Europeans, they remain part of our story. Latin America is not simply conquering Europeans and conquered natives; it is that, but it is also the consequences of European men mating with native women, their offspring. This process results in what would become Latin America's dominant ethnicity, the mestizos. Given that their genes and culture went into the mixing pot that created the mestizos, we should have some understanding of these conquered people.

Before we get into the three civilizations on the American continents, we must not forget that the Americas were as heterogeneous as any continent. Populations ranged from primitive tribes, such as the Yanomamö, which have survived until today, to the large, sophisticated tribal groupings of the Pacific Northwest. There were both tribal and civilized populations in the New World, and their spatial span went from the Fuegians at the tip of South America to the Inuit at the Arctic Circle. The Americas had all types of starting material and environments, but of all the possibilities, only three civilizations developed.

Of the three civilizations in the Americas at the time of discovery, two, the Aztecs and the Incas, were young, as they had been civilized for only a hundred years or so. They were late bloomers, just like other similar advanced populations in sub-Saharan Africa. Only the Mayan civilization had an extensive history. They had peaked by the ninth century CE and were on the downslope of progress when the Spanish arrived.

The Maya were difficult to conquer. They consisted of independent city-states and never came under a central authority. Consequently, the Spanish had to cut off their heads one by one; no single conquest would conquer all. This is in contrast to the Aztecs and Incas, who had a central authority. With them, one clean dispatch was sufficient.

A point we will develop a bit later is that all three civilizations were located in what was to become Latin America. The other part of the New World, the part colonized primarily by the English, had no civilized natives, only tribes. We will later discuss how this difference in human capital influenced the differences we see today between Latin America and its neighbors to the north. The main point is that Latin America resulted from the conquest of civilized people, those having passed the portal of the Urban Revolution.

The New World civilizations had some degree of technology. The maize that existed when the Europeans arrived was so highly domesticated that it had to be propagated by humans; it was not capable of self-propagation.[104] Commerce offers another example. Some "used money, kept accounts, and had sophisticated trading networks."[105]

Generally, they lacked. For some reason, no one in the Americas developed iron, let alone steel, as those in Europe, Asia, and Africa had managed to accomplish.[106] They had one of the most important inventions of humanity, the wheel, but it was simply for toys.[107] Why invent a real wheel when there are no real horses or real oxen to pull whatever is attached? A further point to consider is that they also did not invent the wheelbarrow.

The preceding perspective, while presenting a somewhat underwhelming portrait of American natives, does not mean that all aspects were such. "When the Spanish *conquistadores* entered the Aztec capital of Tenochtitlan in 1519, they found a city larger than Seville, from which many of the Spaniards had come."[108]

Only the Maya had a fully developed writing system.[109] A consequence of illiteracy is that the transmission of culture must be by oral, not written, tradition. Furthermore, this critical lack encourages ideas to stay in one place, a point discussed by Sowell in making an important comparative observation.

The Europeans could steer across the ocean with "rudders invented in China, calculate their position . . . through trigonometry invented in Egypt, using numbers created in India."[110] Their knowledge was "preserved in letters invented by Romans and written on paper invented in China"[111] and their military power "depended on weapons using gunpowder, also invented in Asia."[112]

The foregoing provides a depth of understanding as to why the Europeans conquered the Americans and not the opposite.

New civilization from conquest. The discovery of the then unknown Western Hemisphere by Columbus in 1492 is celebrated by some and vilified by others. Good or evil, it was consequential.

Sowell states, "It was one of the most momentous events in the history of the human race."[113] Additionally, "nothing would ever be the same again, in either half of the world."[114]

Independent of personal view, agnostic, favorable, or not, it was an event of incomparable consequence. Besides the obvious, a consequence of conquest was the creation of a new civilization, the topic for this section.

The end of the fifteenth century saw the unification of the kingdoms of Castile and Aragon, making Spain a major power on the world stage. Operating an empire takes money, especially for the requisite army, and money was what the Spanish king needed to maintain position as a dominant power. The discovery of New World lands abundant with gold and silver was exactly what the king needed. It did not take long for the Spanish to go to the newly discovered lands to collect all that lucre and then ship it back to the needy nation. But first, they had to conquer the locals.

For conquering, the Spanish had guns and steel, while the natives only had sharp obsidian rocks and wood clubs. Adding insult to injury, the locals had to walk to the fight, while the Spanish could ride in grandly on horses. The very few easily defeated the many.

Never before having seen a horse, let alone a man mounted on one, their first encounter with mounted Spaniards must have been terrifying for the natives. The invaders must have appeared as gigantic, indomitable creatures, half man, part beast. Many native warriors must have simply turned and run.

The conquerors had more than just the steel and horses required for a short-term conquest; they did not know it, but they had the weapons guaranteeing their longer-term dominance. They had germs never before exposed in the New World; the first resulted in quick death, the second a slower one, but in the longer term, just as certain.

The encounter between the newly discovered New World and the Old was a catastrophe for the New. Those from the Old World had a long history with zoonotic diseases from their domesticated animals. They had developed a degree of immunity, whereas the previously unexposed people of the New World had absolutely none. On exposure to these imported diseases, they died in huge, genocidal-level numbers. Within 100–150 years of first contact, death from imported diseases had reduced the native population to about 20 percent to 5 percent of its initial level.[115] In modern times, this translates to nine out of every ten houses in a city being vacant.

If the conquering Europeans had any clue about disease and its causes, future generations would probably label this as genocide. However, this label is inappropriate; the best we can do is call them ignorant, which they were. It would take until the middle of the nineteenth century before we had any inkling as to the source of disease.

During colonization, three processes were in play. The first was the importation of Spaniards and Portuguese into the New World. The second was the decimation of the native population by disease. The third, a topic we have briefly discussed, was the genesis of a new ethnicity: the mestizos.

The Spaniards and Portuguese who came were men, men with little interest in settling. For instance, the ratio of female to male immigrants in Mexico was about 1:10.[116] Not only were the new arrivals predominantly men, they were Catholic men. The Catholic Church, which had a say in most matters, wanted no truck with argumentative Protestants; none need apply.

The Spanish were not farmers seeking a new life; they were conquistadores in every sense of the word. They came to take and they came well prepared with weapons and horses. There were no women in their baggage. Being single men far from home, they took to their beds native women, and the resulting offspring, part European and part native, became an intermediary ethnic group,[117] the mestizos. This became a successful biological enterprise, as evidenced by Sowell in pointing out that "in Spanish America, . . . a substantial part of the entire population was of mixed blood after three centuries of rule by the Spaniards."[118] Landes's comments add a final touch to the picture: "The mixed bloods became the overseers, the foremen, the shopkeepers, the petty officials. The Indians were assigned and conscripted to labor in the fields and mines, in homes and on the roads."[119] The mestizo sons became the overseers of their native mothers, aunts, and uncles.

Considering the huge decrease in the native population along with the concurrent influx of people from the Old World and the racial mixing of populations, we must consider Latin Americans as a new civilization, one starting in the sixteenth century.

It is a civilization stratified by "purity of blood." The elite at the very top are the Iberians, as are their pureblood offspring, the creoles. The natives, conquered and with their lands occupied, are the subservient lower, peasant class. In the middle are the mixed-blood mestizos, the middle-class-to-be.

If we are to understand the economic prosperity or its lack for Latin Americans, we must understand the new civilization that arose at the juncture of native insiders and conquering and colonizing outsiders. This is our proper starting point for examining the causes underlying the wealth and poverty of Latin America. What happened before conquest only rises to the level of somewhat interesting prologue.

We have conquering Europeans and conquered civilized natives, and we have the progeny of their mixed matings: three classes. There is an additional source of human capital, sub-Saharan Africans. The Portuguese brought them as slaves into Brazil by the millions, and they too mated with the existing ethnicities. Considering opportunities and possibilities, there are now many more classes.

Humans classify, especially other humans. For good or bad, this is something we do. Back in the day of colonizing the New World, the Spanish created a classification system based on race. However, its roots sprouted from an older concept, "purity of blood," having a meaning a shade different from race.

In the fifteenth century, Spain required that non-Christians convert to Christianity or leave, a law specifically applying to those of Jewish or Islamic heritage. In that period, non-Christians had proscribed rights; only "ethnic" Catholics had full participation in society. Many Jews and Muslims chose to stay and converted; that is, they became *conversos*. They gave up faith in exchange for what they expected to be full social participation; as we will see, they were fooled.

A question arose: was a *converso* equal to an "ethnic" Catholic? Was there still not a difference? This is where "purity of blood" enters the conversation. The answer was that *conversos* lacked this purity of blood; they were not "ethnically" pure. They continued to be denied rights available to "ethnic" Catholics. Proof of purity became a requirement for full social standing, and an industry for creating false genealogies came about.[120]

When the Spanish traveled to the New World, they brought along this idea of purity of blood. In retrospect, they were smart, as this genealogical certification of purity enabled them to remain for a very long time as the dominant elite in Latin America. Lesser people, those of impure or lesser blood, were assigned lower status. They still are.

As expressed in the following quotation, this system was serious stuff. Status did not come from occupation or something similar, as it does now, but from race:

> Created by European (white) elites, the sistema de castas or the so-ciedad de castas was based on the principle that people varied largely due to their birth, color, race, and origin of ethnic types. The system of castas was more than socio-racial classification. It had an effect on every aspect of life, including economics and taxation. Both the Spanish colonial state and the Church required more tax and trib-ute payments from those of lower socio-racial categories. Related to Spanish ideas about purity of blood (which historically also related to its reconquest of Spain from the Moors), the colonists established a caste system in Latin America by which a person's socio-econom-ic status generally correlated with race or racial mix in the known family background, or simply on phenotype (physical appearance) if the family background was unknown. From the colonial period, when the Spanish imposed control, many wealthy persons and high government officials were of peninsular (Iberian) and/or European background, while African or indigenous ancestry, or dark skin, generally was correlated with inferiority and poverty. The "whiter" the heritage a person could claim, the higher in status they could claim; conversely, darker features meant less opportunity.[121]

The classification system in table 12.2 comes from the Wikipedia entry for Casta. It speaks volumes about the classifiers but says

TABLE 12.2   Class System of Colonial Latin America

Parent	B	W	W	N	B
	↓	↓	↓	↓	
1st generation	mulato	criollo	mestizo	zambo	
	↓	↓	↓	↓	
2nd generation (with one white parent)	morisco	criollo	castizo	moreno	
2nd generation (with one native parent)	chino	mestizo	cholo	cambujo	
2nd generation (with one black parent)	negro fino	mulato	cimarrón	prieto	

Note. B = black; W = white or Spanish; N = Amerindian or native.

nothing as to the classified. This new civilization, the Latin American, came color-coded for status.

There is a deeper consequence than simply class founded on color. It is status founded on parentage and not self. The Latin American civilization was founded on the premise that the value of an individual was inherited and never earned by self. "Self-made" is an oxymoron in such a system. Our Differencing Engine runs best in a meritocracy, one where "self-made" exists and flourishes.

In the context of our narrative, this class system is similar to that of the Hindus. Its main consequence is that it impedes mobility, constraining our Differencing Engine. This is one reason Latin America, even though populated by civilized people, performs less well than it should. We will shortly see that there is a second impeding factor: patrimonialism.

Differences between English and Spanish America. The economic outcome of Latin America has been different from that of English America, the United States, and Canada. Two differences offer a partial explanation. First, the English and Spanish had different motives for going to the New World. The other is that they encountered and had to contend with different types of natives, one tribal, the other civilized.

When the English followed the Spanish to the New World, they not only went to a different locale, North America, they left home and made the arduous journey for a different reason. Unlike the Spanish, who came searching for gold and silver, they sought only to make a new life. They were settlers, looking for freedom from the religious conflicts encountered in post-Reformation Europe. They came not as conquerors planning to plunder and take; they were settlers intending to build, and they did. They made communities, extended families, productive farms, and, in time, important cities. Significantly, they brought to the New World a sense of freedom and individuality not comprehended in Catholic Spain or Portugal, and just as importantly, unlike the Iberians, they brought no religious test for admittance. With all being welcome, diversity ensued, strengthening the social fabric to be.

In the New World, the Spanish encountered the formidable civilized nations of the Aztec, Maya, and Inca; the English found only somewhat powerless tribes. To accomplish their ends, the English had only to push these tribes aside and out of their way; brutal conquest, as the Spanish needed with the civilized nations, was not required. The English pushed, shoved, and, when appropriate, bought off the natives with goods from afar. Though it was not completely absent, brute force, especially large scale, found little need or use.

The following contributions from Landes and Sowell help complete our picture:

> Where the English found a land lightly peopled and pushed the natives out of the way to make room for settler families—creating over time an absolute *apartheid*—the Spanish found the most densely populated parts of the New World and chose to intermarry with the inhabitants.[122]

> The colonies of transplanted British and French communities in North America remained a largely insular world of their own, with Europeans living in European communities and Indians living in Indian communities.[123]

The key word is *apartheid*; the English did not mix or mate, at least to any extent, with the natives. Even after encounter and setting aside, the English and the natives remained separate; they did not conjoin to form any mix of populations—there were no mesti-zos in North America. What we currently have in the populations of the United States and Canada are members of Western Europe.

The Spanish encountered different conditions. Between them and their goal of gold and silver were all three civilizations of the Americas. These were not simple, "shove-aside" tribes, like those encountered by the English. They were sophisticated, civilized na-tions, most with central authority and substantial armies. As an impediment to achieving goals, they had to be forcefully removed, and they were.

There is a bottom-line difference between Latin Americans and their northern neighbors. The natives of Latin America have played a significant role in the region's economic development, both as source material for the mestizos and in their own right. The natives to the north were shoved aside and placed in reservations, never to have a significant role, political or economic.

Spain, from the time of the Roman Empire, had its own feudal system, the *encomienda*, where those in power protected those not and, in return, received forced labor and tribute. Spain exported this dependency relation institution to the New World. There, an *encomienda* was a reward to the elite for valuable service to the Crown and came with "ownership" of several natives from a given community.

The *encomienda* required its grantee to provide protection to his natives. This included an obligation to provide instruction in the Catholic faith. The natives owed him tribute, such as agricul-tural products. Importantly, the *encomienda* required the natives to labor for their owner.

This was feudalism, security in exchange for labor and food, but by a Spanish name. The net result was that the natives were not only conquered people living in an occupied land; they were serfs or, effectively, slaves to the ruling foreigners. A system of forced servitude that was approaching its end in the Old World by the tenth century started afresh in the New World in the sixteenth, enduring until abolished in the late eighteenth century.[124]

Sowell, in commenting on the relationship between the con-quered indigenous people and the conquering Europeans in Latin

America, stated that "seldom was there a situation such as that in Spanish America, where a European overlord class lived directly off work of indigenous peoples living in European-controlled territory."[125]

The *encomienda* adds to the reasons explaining the resultant economic difference between English and Spanish America. In Latin America, the economy commenced with slave labor.

Patrimonial Latin America. To understand why Latin American economies perform below potential, there is one last trait to examine. They have a patrimonial bent; this is where those in power dole out goodies to friends and relatives and in return get help in maintaining power. Those not belonging to a family or power clique get no piece of the economic pie. The "purebloods" keep it all for themselves, while those not belonging to this in-group remain economically repressed. Given the nature of the hierarchal structure of Latin American societies, this is both natural and predictable.

The foregoing definition of *patrimonial* only defines "what" and contains no idea as to "why." Fukuyama, in *The Origins of Political Order,* offers a definition with greater utility and, in doing so, takes a very important step. He couples the political act to a biological imperative by tying patrimonialism to both kin selection or inclusive fitness and reciprocal altruism. This connection provides the why, and once we know why, we can predict.

According to Fukuyama, patrimonialism has a basis in reciprocal altruism and kin selection. It is what comes naturally:

> What I have labeled patrimonialism is political recruitment based on either of these two principles. Thus, when bureaucratic offices were filled with kinsmen of rulers at the end of the Han Dynasty in China, when Janissaries wanted their sons to enter the corps, or when offices were sold as heritable property in ancient regime France, a natural patrimonial principle was simply reasserting itself.[126]

In kin selection, we act to help our relatives; they share our genes and helping them helps our genes. This is the foundation of the gene-centric or "selfish gene" view of human nature. Would it be unexpected for the head of a Latin American country to make a relative the head of customs, a position of low pay but huge self-enriching potential?

Reciprocal altruism is different. It is of the form "I will scratch your back if at some future time you will scratch mine." It is a consequence of always living in social groups, where such behavior promotes group adhesion, a characteristic of groups in group selection. The key difference is that reciprocal back scratching is independent of genetic relatedness. The consequence is that human nature is to assist those who belong to our group, even if they are not genetically related.

When Latin America created its class structures, it created an environment nourishing a patrimonial lubricated polity. In this environment, each racial or "blood" class of the social hierarchy is a group. Because of kin selection and reciprocal altruism, the individuals of the ruling class act to keep power in their group through patrimonial behavior. Their allegiance is to their group and not to the greater society of the nation. Creoles took care of the creoles; mestizos and natives need not apply.

A class-based hierarchy is not a requirement for patrimonialism. In any nonegalitarian society where there are elite, they tend to act as a cohesive group protecting their own interests. The class-based structure of Latin America just made the process "more natural" by providing lines of natural cleavage. The Chinese, with the Imperial Examination, prevented this natural patrimonial proclivity of their elite, effectively preventing them from acquiring power that could contend with that of the Emperor.

A consequence of a patrimonial base polity is restriction of economic activity. An individual who would otherwise be motivated to improve life's condition will not, because such systems *a priori* block success. Why try when the patrimonial system guarantees failure?

Another consequence is that patrimonialism reserves economic activity for the elite. This might be good for the elite, but it is bad for the nation. Consumers boost the nation's economy, but nonplayers in the nation's economy cannot consume; they have no money. By keeping the economic pie small, even the elites lose.

The biggest loser in a patrimonial system is the greater society. Its Differencing Engine only works for the elite; the rest remain unchanging, frozen in place. Remember, there is no resultant change in a population's genome or culture when the economic rewards come from heritable surnames and not heritable traits. The ultimate result is no changes leading to improvement in the economic acumen in the nonelite portion of the population.

Landes, as usual, nicely expresses the consequences:

> This pattern of arrested development reflects the tenacious resistance of old ways, and vested interests. In particular, the apparently rational focus on land and pastoralism (long live comparative advantage!), reinforced by social and political privilege, bred powerful, reactionary elites ill-suited and hostile to an industrial world.[127]

Patrimonialism is not just a problem of Latin America. Tribally based societies also provide lines of natural cleavage that promote patrimonial behavior. This has been a continuing problem in sub-Saharan Africa, which we will cover next.

Improving Latin America's economic prospects. Latin America, just as Russia, could have higher economic output, and, similar to Russia, part, if not most, of the solution will be by changing governance. Latin Americans appear to have moved past the banana republic (with the military dictator) stage to ones that are more democratic. They still elect socialists and populists, suggesting a lack of political maturity. As discussed, patrimonialism remains a persistent problem. However, they are getting there.

One problem some Latin American nations face is significant indigenous populations. Like other similar populations, theirs do not perform well economically, and, where they have substantial numbers, they are a strong political force. Typically, they have been a force for socialist programs, a force *contra* liberal democracy. States wanting liberal democracies need focused inclusive programs that will bring these populations into the modern economy to share the wealth.

Sub-Saharan Africans

Introduction. A point that we have made throughout our narrative is that populations not transitioning through the portal of the Urban Revolution and becoming civilized are the poor of today. When we earlier compared the economic performances of our eight groups, sub-Saharan Africans were at the poor end of the economic spectrum. They were the poorest, and the question arises, does our thesis hold true?

In the year 2000 BCE, what was to become China had three thousand polities. A bit less than two millennia later, there was one; China was civilized. England was similar. At the start of its conquest by Rome in 43 CE, thirty Celtic tribes populated it. Between then and the Norman Conquest in 1066, other groups, such as the Angles, Jutes, and Saxons, had entered the pot, and now, all we see are Brits. They are a single people, and finding a Jute, an Angle, or even a member of one of the original tribes would take serious scratching. Clearly the civilizing process unifies multiple ethnicities into a people with a common culture.

When we have discussed changing identity from tribe to state, this has been the consequence about which we were talking. Building cities, forming states, and creating civilization is a process where the many become one. Unlike China or England, Africa has thousands of ethnic groups, a difference suggesting that, by our definition, it is not civilized.

We can use another broad-brushed test, one based on Childe's list of ten characteristics of the Urban Revolution. Two not found in sub-Saharan Africa are as follows:

1. A characteristic of the foundational urban city was large-scale public works represented by monumental public buildings.

2. Management of the urban enterprise required written records for such things as size of harvest and the amount of tax due. It also required numeracy for accounting. Some level of literacy and numeracy is a characteristic of the urban enterprise.

Ruins of past major human habitations such as those of the Great Zimbabwe exist in sub-Saharan Africa, but the big tokens of civilization, truly monumental public building, such as pyramids, stately buildings, or temples, do not. Likewise, there is no evidence of literacy or numeracy.

The absence of these two characteristics does not necessarily mean that sub-Saharan Africa had no states or civilizations of any sort. The world is not black and white; it is a shady gray, often blurry. At the start of the Age of Discovery in the late fifteenth century, there were cities, states, and civilizations, just not very many, advanced, or developed. Just as important, sub-Saharan Africa had large swaths of land containing people outside any umbrella of a developing civilization. It was a subcontinent transitioning only in selected parts to the urban world and to civilization. It was a world where only some few people had started the transition, others not. This is the land encountered by the European colonizers.

The state and precolonial Africa. Environmental determinism explains some of what precolonial sub-Saharan Africa was not. The central issue is that it is a hard place to get around. Its coasts offer few sheltering harbors. It has rivers, but they typically have poor navigability. Its torrential, tropical deluges play havoc with dirt roads, and, in some areas, forests quickly reclaim roads. Where the tsetse fly lives, horses for transport cannot; travel is by foot. All the foregoing stymies activities requiring communication, the consequence of which cascades into multiple sectors of human development. It impedes on multiple fronts.

When we discussed the West, we learned that a reason for cities, other than as administrative centers, is trade and commerce. Logically, cities are founded on trade routes, and trade routes infer navigable waterways and roads. Difficult communication makes city development problematic. This is the case for sub-Saharan Africa.

As discussed, states and civilizations have as a starting point the city. Urbanization is important because that is where the state is strongest and, in some cases, the only place where it exists.[128] If city formation is restrained, so is the formation of states and any possible ensuing civilization. The difficulty of communication within sub-Saharan Africa predicts fewer and less advanced states and civilizations when compared to lands less restricted.

There is a second aspect to poor communications. A state, once founded, has a central authority, and it can exist only in those areas where it can project that authority. This means that in all areas claimed by the state, it must be able to communicate its authority. The Ashanti nation of West Africa, which at its peak approximated present-day Ghana, provides us an example as they "conceived of their empire as radiating out in all directions for twenty days walking."[129]

This concept defines the state's territory as a circle having its principal city at the center and with a radius equal to the distance that could managed. For the Ashanti, that was how far a person could walk in twenty days. In the savannah of West Africa, where horses and camels were usable, attainable distances were longer. In this zone, the radius of authority was considerably larger, leading to bigger states.[130]

With the foregoing, we can create a simple model of the physical state of precolonial sub-Saharan Africa. For various reasons, urban centers form, and, with them, so do states. This makes the state's central power resident at an urban center. This power diminishes with distance from authority, and at some distance, a twenty-day walk for the Ashanti, the state ceases. Outside of these "circles of state authority" are the hinterlands, the lands of the stateless and uncivilized.

These uncivilized hinterlands are important to sub-Saharan Africa's story. In addition to these areas not being under state control, other important, related factors are in play. The hinterlands tended to be inhospitable, resulting in low population density,[131] and the people who lived there were primarily subsistence-level agriculturists, having consequences for everything from state formation to slavery.

Urbanization and state building require food production in excess of the needs of the producers. This is exactly what subsistence agriculturists do not do; they only produce enough to fill their own bellies and no more. "Selfish" subsistence agriculture is of no value to the state; there is no money in it for the rulers.

The subsistence agriculturists present in precolonial times had two social forms. Both were communal in nature; the superorganismic kinship group was one, while in the other, the extended family was the basis. This latter is often referred to as the familial mode of production. Both are self-contained, and neither produced anything of significance in excess of communal needs. They are effective counterpoints to our Differencing Engine, meaning that economic acumen does not improve in these societies. By definition, they simply subsist.

These subsistence agriculturists can be of economic value; it just takes state coercion. If the state imposes taxes, people are forced to exit the subsistence mode and produce excess for sale. If the people have no alternatives, this is what normally happens. However, sub-Saharan Africa's low population density created an alternative. Plenty of unoccupied land was available to farm, and people could simply move out from under the state's coercive thumb. As an example, "migration to escape from social or political problems was . . . common among the Yoruba, the Edo, the Fon, and many others."[132]

States wanting to expand often do so for the reason of increased revenue; expansion normally increases their tax base. Expanding the state requires investment in the tools of expansion, usually an army. A second point to be considered is that once expansion is complete, there will be an additional cost for its administration.

If the state is considered in the same context as a commercial enterprise, it must make returns on its investment to survive. Unless a new land has natural resources, such as timber or minerals, any investment return must come from its occupants, the peasants.

Subsistence farmers produce only for self, and as they produce nothing in excess of needs, they have no apparent economic value to the state. However, in sub-Saharan Africa, people themselves had value. They were valued not for what they produced but as slaves:

> The point of war was to take women, cattle, and slaves. Thus the slave trade, especially in the eighteenth century, should be seen as part of the process by which African states grew: by capturing people rather than by gaining control over territory.[133]

This is sort of a hidden reason for the lack of state formation. It probably did not take a ruler too long to figure out that capturing land easily abandoned by its occupants was a risky investment. Capturing people for sale as slaves made better economic sense, and that is what they did. As evidenced in the prior quotation, it was an economic decision.

Moving away for purposes of relocating from a coercive state will have a cost if there has been significant investment in infrastructure. For agricultural land, this is often in the form of irrigation systems. Because agriculture in sub-Saharan Africa depended on rain, there was no cost in leaving. People were literally free to leave, because when they moved, they left nothing of value behind.

The result of the foregoing is a condition unknown elsewhere: unimportant land with unimportant borders. Who cares? A consequence of ambiguous borders is that, for a given area, more than a single polity could hold power.[134] As Herbst pointed out, "in some areas of Africa, it was so difficult to broadcast power that there were, essentially, no centralized political organizations above village level."[135] The village as the dominate polity is a token of statelessness and lack of civilization.

Nicolas Wade, in *Troublesome Inheritance,* presents a good overview that shows how sub-Saharan Africa fits into the scheme of civilizations:

> The evolution of human social behavior was thus different and largely or entirely independent on each continent. States had developed in the Middle East, in India and in China by about 5,000 years ago and in Central and South America by about 1,000 years ago. For lack of good soils, favorable climate, navigable rivers, and population pressure, Africa remained a continent of chiefdoms and incipient empires. In Australia people reached the tribal level but without developing agriculture; their technology remained that of the Stone Age into modern times.[136]

My visual image of precolonial, sub-Saharan Africa is a land with scattered areas of urbanization and states. In between these scatterings are the hinterlands. They contain just villages of subsistence agriculturalists and separate the relatively sparse underpinnings of civilization.

This, I believe, explains poverty in sub-Saharan Africa. Before colonization, there were some cities and civilizations where the Differencing Engine could run, but it was outside of these relatively sparse urban zones where the majority lived. Many, if not most, in these hinterlands were subsistence agriculturists. These people are not part of any Differencing Engine, and their economic acumen is stuck in neutral. It will not go forward until they shift their mode of production away from subsistence agriculture.

Sub-Saharan Africa's Population Problem

Identifying the problem. We made the point in chapter 3 that the bigger problem of "the poor" is not poverty *per se* but rather the high population growth associated with it. Admittedly, that statement would be a hard sell to the starving peasant, but revisit figure 3.3, "Growth of World Population," and the evidence is plain. A finite earth cannot support the consequence of such high growth; the curve absolutely must bend to the right. The primary reason for the growth curve going almost straight up is the high fertility of poor sub-Saharan African nations. This, then, not poverty, is the ultimate problem.

In an earlier discussion, we pointed out that the model of the Demographic Transition offered salvation. In this model, once nations become "rich," fertility rates decrease. We may not know why, but they do, and as long as this occurs, we do not actually have to know the reason. Therefore all we have to do to solve the problem is to get the sub-Saharan Africans moving toward the prosperous side of the economic spectrum.

Like some other topics in our narrative, what appears obvious at first sight might not be. This is such a case. Reducing population growth in sub-Saharan Africa might not be simply a case of solving the poverty issue. As we pointed out earlier,

> a conclusion based on figure 3.6 is that high fertility is primarily a problem of poor countries. This is true and, based on our resent discussion about the Demographic Transition, expected. However, there is more to the story. Of the 144 nations shown in figure 3.6, 41 have a fertility rate greater than 3.5, and of those, 35 are sub-Saharan African.

This observation raises the possibility of other issues at play. There are, but they are esoteric and not part of our central narrative. We are going to reserve that discussion for the last chapter.

Independent of cause, a consequence of the foregoing is that a poor nation with a high fertility rate reaching a Stage 4 level of the Demographic Transition might not have its population growth curve flatten out. Therefore solving the poor problem might not solve the population problem. However, we will not know until the poor nations have in fact reached the prosperous stages.

Therefore the real question we need to ask is not about poverty but about population growth rates. Why is it that poor sub-Saharan Africans have such high birthrates while most similarly poor nations do not? That is the topic for this section.

Who are these over-reproducers? If we assume Kenya to be a representation of the greater sub-Saharan Africa, the problem lies with the rural population; that is where the large majority of the population lives. Kenya's 1962 census showed that only about 7 percent lived in urban areas,[137] with most making a living from agriculture. However, changes were under way, and by 1979, the urban population had increased to 15 percent, reflecting the large movement toward urbanization.[138]

The problem is rural, but it is also more. From 1962 to 1978, Kenya's population grew at a rate of 3.7 percent per year, the highest rate in the world. In 1962, its population was 8,636,266, which, by 1978, a period of only sixteen years, had doubled to 15,327,061.[139] There are two questions: why all the babies? Why all the extra babies? In answering these questions, we need to understand their relation to the rural population.

The first question we need to address is, where did all the extra babies come from? For this, we can assign blame to education; it was an unintended consequence. It also makes for an informative story.

The story starts with a high fertility rate but a high infant mortality. In 1962, the fertility rate in Kenya was 6.8 children per woman in 1962, rising to just under eight by 1979.[140] These are huge numbers. Simply consider that for 2015, the highest estimate is for Niger at 6.76,[141] whereas for France, it is 2.08. Couple this very high fertility rate with an initially high infant mortality rate decreasing from 119 to 83 per 1,000, or 30 percent, over the same period,[142] and the result is obvious: a lot more Kenyans.

We can find part of our answer in the reason for the decrease in infant mortality. Over the period in question, education reached the hinterlands. Girls learned about health, hygiene, and nutrition, and when they became mothers, their children had better survival prospects.[143] Education of the previously uneducated saved babies. The good of education resulted in the bad of too many people.

Now we know some about why all the extra babies; it is time to understand why so many in the first place. This part of the story is a bit more complicated and is where we tie the problem to the rural environment. Who are these people with rapidly expanding populations? What is their nature? We are going to start by examining how they make a living or "mode of production," a topic covered in the following quotation:

> The first model is one of a mode of production made up of what, for want of a better term may be called "communal cultivators." The second represents a mode based upon peasant farmers. The parallel parts of these models are set out below. It must be emphasized that, like all their kind, they are abstractions and simplifications, particularly unlikely to exist in their pure form in Western Africa. The two processes with which we will be concerned are closely related to them, since one is the process of change from communal cultivator to peasant and the other the process of the incorporation of communal and peasant societies into the world capitalist network. It must be stressed that both these processes have been occurring simultaneously in Africa in the last hundred years.
>
> Communal cultivator
> (a) Communal land ownership: group or individual land use.
> (b) Social division of labor largely based on kinship.
> (c) Markets absent or peripheral.
> (d) Political hierarchy and obligations largely conterminous with kinship.
> (e) Largely homogeneous culture.
> Peasant
> (a) Individual land ownership: group or individual land use.
> (b) Separation between social division of labor and kinship.
> (c) Market principle in operation.
> (d) Separation between political hierarchy and obligations and

kinship.
(e) Distinction between "great" and "little" cultures.[144]

The "communal cultivator" appears to be our superorganism, or very similar to it. Individuals work on behalf of the community, and the community in turn takes care of the individual. It is subsistence agriculture, and the unit of production is the community.

The "peasant" is highly similar to the "communal cultivator." What changes is the unit of labor, which is now the extended family. This is the familial mode of production.

There is a third type implicit in the foregoing. It is the farmer as a unit of the "world capitalist network." This is where agricultural activity changes from production for consumption to production for gain or profit.

There is a critical observation, one where we find an important clue to our question:

> High fertility is enjoined by the familial mode of production, which is associated with peasant economies, hunting and gathering, shifting cultivation, and nomadic herding. Under the familial mode of production, the head of the family can benefit from exploiting the labor of other family members, and thereby stands to gain from ensuring high fertility within the family. When the capitalist mode of production has been firmly established in these societies, however, the family has to invest in the offspring to make them productive members of the new economy; this reverses the intergenerational flow of wealth from going up the age hierarchy to down the age hierarchy. At this point, the society turns around completely, and childbearing is no longer economically beneficial.[145]

In the familial mode of production, children have economic value. On the family farm, they produce more than they consume; they are a net economic asset. Besides, they are the vehicle for getting genes down the road. Families making a living in this fashion want more children; they help provide survival security. This, I propose, is an important driver of sub-Saharan Africa's growth rate.

The foregoing ties in with an observation we noted earlier, that of the massacre at Crow Creek, where absent at the site were the remains of women and young children. Our discussion was about access to women and war; we did not discuss the children. It could be that children were taken and not killed because of their economic importance; they became extras hands for working the garden. Considering people in solely economic terms might not be comfortable, but it reflects reality.

In the world of the West, children are blessings and are not often thought of within an economic context. If they were, they would plainly be an economic drag. They do not produce more than they consume; they add nothing to the family coffer. According to the U.S. Department of Agriculture,[146] the cost to raise one child to age eighteen is just north of a quarter million dollars. This economic reality probably has a role in developed nations having flat to slightly decreasing populations.

Improving sub-Saharan Africa. Familial mode of production is the problem, and the following quotation lends insight into its nature in sub-Saharan Africa:

> Most people are still tied to land within extended family units, and the major means of production (i.e. land) is largely owned communally and by smallholders rather than by an ownership class. At the same time, the state in league with other foreign and local elites, who dominate the process of capital accumulation, have [*sic*] helped to perpetuate the familial mode of production to further their own political and economic interests. The results are that the forces of production (i.e. land, labor, and capital), while certainly influenced by market forces, are often allocated on the basis of nonmarket criteria such as kinship obligation or custom whether real, invented, or politically motivated.[147]

As long as the family or commune controls the land needed for subsistence farming, and lacking better alternatives, the familial mode of production will continue. The state could alter land control, but the political and social consequences would be too high— people would revolt. A better alternative is needed. It has to be more than simply an alternative to survival at some base level; that is what they currently have. It has to be an actual improvement.

Solving the population problem involves how people have been living for a very long time and will be difficult. We earlier discussed the development economist Willian Easterly and his "searchers," repeated here:

> Easterly's searchers are the bottom-up people who address narrowly

defined, specific needs. They make no assumptions about people being the same or different. By working directly with the people, at the level of the individual, they do not have to make any assumptions. Actual behavior, whatever it is, is naturally baked into the solution.

The searchers represent a small-ball process, one addressing problems at the level of the individual. They do not address large, society-wide issues. The model is to incrementally solve small problems at the lowest level and the broader ones at higher levels should take care of themselves. This is the process Easterly advocates, as in his view, it has a better chance to succeed than the processes put forth by the planners.

A problem with Easterly's searchers model is that it has an underlying assumption of commonality of motive. Some populations have social institutions affecting labor for economic ends. In these societies, labor has a strong social component, one confounding what Easterly sees as primarily an economic problem.

Economists are right in stating that people will do what is in their self-interest. What makes Easterly's searcher approach attractive is its black-box nature: there is no need to understand why. It works because individuals do what they perceive as best for their survival. It is sort of like making Mother Nature a co-conspirator.

CHAPTER 13

Lagniappe

◇◇◇◇◇◇◇◇◇◇◇◇◇◇◇◇◇◇◇◇◇◇◇◇

By the end of the previous chapter, we had answered our question, why are some nations rich while others remain poor? and we could have ended our narrative at that point. Logically, given this chapter's location, "Conclusions" could have been its title, but, given what we will next cover, "Consequences" might be more appropriate. Although we will discuss consequences, there is more, and that too does not quite fit the bill.

In New Orleans, where I live, there is a word of seemingly French origin used by both Cajuns and Creoles, *lagniappe*. It came to south Louisiana not from the French but by the initial colonizers, the Spanish. They had borrowed it from Quechuan, the primary language of the Incan Empire. It means a little something extra, sort of a baker's dozen across the Sabine River in Texas. The *amuse-bouche* you get at a high-end restaurant is a *lagniappe*; it is a nice, but not necessary, extra. With this last chapter, we get such a little extra.

The following topics are not needed but nice to have—topics for *lagniappe*.

1. *Minority populations.* We have performed our analysis using groups of nations, and it should be straightforward to use our conclusions directly to make between-nation comparisons. Can we also use our analysis to understand mi-

nority populations? For example, would it be proper to use it to understand Latin Americans living in the United States or recent Muslim immigrants in Germany? That will be our first topic.

2. *Out of poverty.* Urbanization was the path to riches for some. Is it still the path out of poverty? There are data we have yet to examine that will be useful in answering this question.

3. *Reality and consequence.* Our narrative develops the steps leading to our current reality. This section presents the consequence of that reality.

4. *Killing biology.* By minimizing war, solving famine, and coming close to defeating disease, we have learned how to avoid the biological realities all other organisms must endure. Because of our technology, we are all equally fit. This makes survival of the fittest for us humans a *non sequitur,* marking the end of human Darwinian evolution as well as our Differencing Engine. Our story ends.

Minority Populations

Our study explains the difference in relative economic performance between religions/geographic areas, and essentially, it is about differences between nations. The next level down the hierarchy for consideration is minorities living among a majority host population. Examples are Muslims in Paris, overseas Chinese in Malaysia, and Parsee in India.

These paired groupings present questions similar to the one asked in our narrative, and some will be tempted to use our study directly to answer them. Our study addressed a specific question, making its use for different ones a risk. Its unwitting use could easily lead to wrongheaded conclusions, and, as a prophylactic to such misuse, it will pay to examine these issues.

"Can we extend our study and use it for other than its original quest?" is the question for this section. I believe we can, provided it is only part of a broader approach, one addressing requisite caveats. The hard part, of course, is determining what these caveats are, and in this section, we will address some.

These paired groupings of minority–majority have multiple flavors. One, covered in depth by Amy Chou in *World on Fire: How Exporting Free Market Democracy Breeds Ethnic Hatred and Global Instability*, is the case where there is an economically dominant minority living within a poor majority. Examples are the Chinese in Southeast Asia, Lebanese in West Africa, Indians in Fiji, and individuals from the West in Latin America. Economically successful Jews in Eastern Europe are similar to this grouping, except for the majority population being middle class, not poor.

On the other side of the coin are minority populations from a religion or region we have classified as poor living within a rich majority. Examples are African Americans in the United States and Muslims in Europe. Hispanics in the United States are similar, but they are from middle-class Latin America. In their case, the difference is a matter of degree.

There are cases where the minority is from a rich religion or area and is a minority in a likewise rich nation. Koreans in the United States are an example.

Cases we have not considered in our study are indigenous people in a conquered land. Cases are the Aborigines of Australia and Native Americans in Latin America, the United States, and Canada. The Malays in Singapore are also an example. These are poor minorities living among rich majorities. In certain cases, an indigenous group might actually be the majority in the specific land they occupy. The Guajiro people of the Guajiro Peninsula of Venezuela and Columbia are an example.

Not all cases fit into this classification scheme. Sometimes the minority comes from within. One example is the Igbo tribe of Nigeria, where they are a dominant minority among other similar tribes.

Clearly the first caveat for using our study is that each member of the pair must be from one of our eight classifications. Native Americans, Igbo, and such are outside this bound. A minority from a Muslim nation living in a Western one would fit the bill.

General cases. Earlier, we mentioned two general cases having primary interest. One is a minority population from a poor nation living in a rich nation, and the other is the opposite, a minority population from a rich nation living in a poor one. For each, we want to predict the relative economic performance after one year and again after one hundred years.

Based on our study and without other factors, we should expect the immigrants from the poor group to have lesser economic outcomes than their rich hosts. They possess lesser economic skills and intelligence when compared to their hosts. This what our study shows, and, no matter what the state does to compensate, they will have poorer economic performance. Their economic performance, as a group, will be on the left-hand side of the distribution curve. This is the unvarnished truth.

This result certainly applies for the one-year scenario. Unless there is reason for the type of change we have discussed throughout our study, we should expect similar results after a hundred years. However, because culture copies, we might expect the economic difference to be less than that at one year.

If we look at the scenario of the rich group as a minority in a poor nation, they should have superior economic results. This also applies to the one-year and hundred-year scenarios. Predicting otherwise for either scenario is contrary to our narrative. Having a different outcome would require change, and nowhere in these two scenarios do we see change or even a reason for change.

The foregoing suggests that we expect the relative economic outcomes to be deterministic, and given the stated conditions, they probably are. People carry with them wherever they travel their genes and cultures. They carry their abilities and inabilities. Even over a hundred-year period, there is insufficient time for genetic change; allele frequencies will not have time to shift to any major extent. One variable, culture, might change. However, when we look at similar real-life situations, those from poor cultures are slow to adopt cultures of rich ones.

Regarding poor majorities with economically successful minorities living in their midst, Nicholas Wade makes the following commentary:

The Malay, Thai, or Indonesian populations who have prosperous Chinese populations in their midst might envy the Chinese success but are strangely unable to copy it. People are highly imitative, and if Chinese business success were purely cultural, everyone should find it easy to adopt the same methods. This is not the case because social behavior, of Chinese and others, is genetically shaped.[1]

I mainly agree with this statement, but I wish to take issue with the binary nature of Wade's argument. He presents a case that it is either genetically shaped behavior or one that is purely cultural. The correcting point I want to make is that it is probably both. However, the main point is that even after centuries of differential successes by the Chinese, Southeast Asians remain economically maladroit.

Consideration of population variance. In making comparisons between a minority group and its host majority, there is often an unstated assumption: the minority population is an unbiased representative of its original native population. It is only when this minority population is from such a sample that we can use our study and make valid comparisons.

In our study, we placed the people of the world into eight bins, and if we were to parse all eight, the number of identifiable groups would reach into the multiple thousands. There are far too many to make it practical to enumerate. Importantly, there is no such person as a typical Latin American or a typical anything else. For our analysis, we lumped together people of various stripes, and we need to know which stripes apply when we address minority populations.

Any analysis would be simpler if the minority group in question represented a random sample of their larger population, but this is almost never the case. Even with the huge recent migrations from Latin America to the United States, these people are not from all over, not the so-called random sample. They are mainly from Mexico and Central America, and any from nations like Uruguay or Venezuela are probably rare.

This problem of being a nonrepresentative sample is probably the largest caveat in using our study.

History of the minority population. We need to consider the history of the group in question. The Parsee minority of India originated in what is now Iran, which they left about a millennium ago to avoid persecution by Muslims. Not only have they been resident as a minority population for a very long time but they are a highly endogamous society and have not mixed genetically. Populations seem to maintain their identities, and in cases like the Parsee, assimilation might never occur. Over the intervening years, they might have forgotten the dust of the old country, but they have carried with them into the present their old cultural ways, along with their genes.

With their long time as a minority population, African Americans are similar. Even though their time as a minority is not as substantial as the Parsee, time away from origins and their history over that time have probably resulted in changes, making our study problematic for this group.

Minority populations only recently distant from their origins probably reflect their origins and the character imputed in our study. A case would be recent Middle Eastern Muslim immigrants in Europe. Being fresh from their origins, their history should not warrant special consideration.

Other consequential factors. Factors other than being a non-representative sample or having a history apart from the original population can result in our study not being applicable. A minority population becoming assimilated or not is one such factor. Clearly if a minority population mixes and mates with the host, after a time, evidence of the minority ceases to exist. At the opposite end, if the minority population is fully endogamous and maintains all its old ways, its difference to the host will be fully maintained.

An example of complete assimilation is Western Europeans in the late nineteenth and early twentieth centuries immigrating to the United States. Because the United States was mainly a mixture of Western Europeans, the evidence of the more recent immigrants as a minority does not exist. At the other end of the spectrum are the Jews in Eastern Europe. They were strongly endogamous, not mixing with the majority population. They remained as a distinct minority.

Another factor relates to a change of institutions. Minority populations typically experience institutions different from their origin, possibly resulting in change. A new institution will not be the cause of change in people but rather a cause of change in their outcomes. Inclusive institutions can unmask latent economic abilities. A group whose economic performance was poor in an original extractive nation might find positive change with a move to a more inclusive environment.

In a similar vein, I suspect that some of the success Indian migrants experience as a minority population is due to the change to a society without a caste system. Examples are Indians as dominant minorities in East Africa and Fiji. Has their improved economic status been a result of moving from a caste-based society? This would be an interesting topic for research.

The mere fact of a minority population's existence can possibly elicit change. For example, members of the host population might not trust their own in commercial transactions, especially if there is a history of conniving or such. They might find that strangers, a minority population, with high internal trust might be better as commercial intermediaries. This could be a factor in some cases of dominant minorities.

Conclusion. It is possible to use our study for other than its original intent, but such use has caveats needing attention. The preceding discussion only exposes some, certainly not all, of the pertinent factors.

There is a larger conclusion. The rich of our narrative are the consequence of the Differencing Engine, which resulted in their improved economic acumen and intelligence, a change requiring about a thousand years. Those not having the Differencing Engine in their history did not make similar changes; they are the poor, and moving to a new land is not going to change that truth. For a given environment, the relative economic performance between the two classes is fixed; those from a rich group will economically outperform another from a poor group. That is true now, and this difference will remain essentially the same, even after a hundred years.

Out of Poverty

Solving the problem of poverty is outside the scope of our narrative, but data of the type we have been using offer suggestions to poverty's solution. Since it fits in so nicely, it will pay to spend a few words. Besides, it makes our story tidier.

When we look back at our narrative, we see that the turning point for economic prosperity was the Urban Revolution. That not all have become rich suggests that perhaps that revolution is incomplete and still in progress. Figure 13.1, showing that a low percentage of urbanization correlates with poverty and a high percentage with prosperity, is consistent with that premise. The Urban Revolution remains a work in progress.

This book has been long enough, but without graphs, it could have been significantly longer. There is, of course, the adage that a picture is worth a thousand words. Graphs, as sorted and organized pictures, are again a thousandfold more informative, and figure 13.1, "Urbanization and National Prosperity," is such an example. It is a picture of our theme song: "those transitioning through the portals of the Urban Revolution became rich while those not remained poor." Figure 13.1 shows exactly that: rich nations are urbanized, poor ones not.

There is more to take away from this graph. One statement it makes is that the problem of the poor is primarily a problem of the rural environment. It further suggests that governments and their agencies have two problems to solve. One is the need to create economic opportunities in the urban environment for enticing the rural poor; the other is to solve the problem of poverty itself in the rural environment—a separate issue.

Poor people moving to the cities solves nothing. The *favelas* of Rio de Janeiro and the shanties of Caracas are testament. There must be employment, and for this, the state has ultimate responsibility. I would suggest that the state could better spend available resources for the poor, not by providing welfare, but by alternatively making investments resulting in jobs. Jobs not only feed bellies but provide dignity, a need as universal as food. Dignity does not come with handouts.

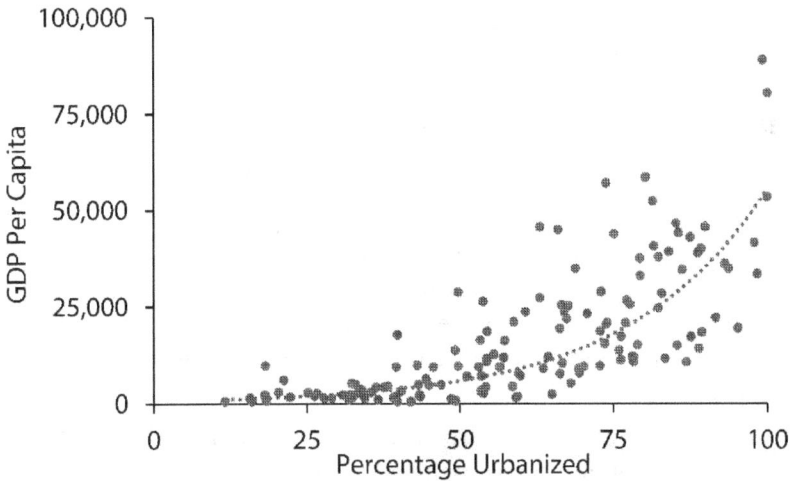

FIGURE 13.1   Urbanization and National Prosperity

I suspect that we know how to do the first, and, in any event, it appears to be a process under way; urbanization is increasing in poor nations. The second, the problem of poverty in the rural environment, is one for which I doubt we have a clue to its solution. In the rural areas of many poor nations, the familial mode of production and/or subsistence agriculture is the means of survival. Both have their roots deep in our kinship group past, when we lived in superorganismic groups. I doubt that aid agencies and developmental economists have much knowledge about such groups and their people—a possible reason aid has not been effective in some parts of the world, especially sub-Saharan Africa.

This section is not about minimizing the problem of poverty in urban environments; as figure 13.1 clearly shows, it is there. It is just not as basic a problem as that of the rural world.

When we look back over our narrative, there is an important conclusion. In it, those populations that became significantly better at economic activities or became smarter took many centuries to arise. There are important consequences to this "slow change"; those providing assistance to the poor must manage with the here and now of reality. They must live with the way people are now and not hang around waiting for needed change; it might not come in any reasonable time.

Reality and Consequence

A graph we developed earlier, figure 11.3, "Wealth and the IQ of Nations," given again here, is a picture of where the world is today. It is our story's conclusion and shows the difference in the economic performance and intelligence between our eight groups. We opened our story with the premise that there is a reason for everything, and our narrative, with its Differencing Engine, provides that reason.

Our opening assumption was that at the start of the Neolithic, we were highly similar in behavior. During that period and into the start of the Industrial Revolution is when the Differencing Engine operated for some, resulting in the differences we observe.

Natural selection operating on phenotypes, culture, and their interactions underlies this Differencing Engine. How much of the difference can we attribute to genes and how much to culture? We do not know. Because IQ has a genetic component, accounting for about 50 percent of its variation, there is probably a significant genetic component.

In chapter 11, we discussed IQ and its significance. The subsection's title, "Intelligence Is a Tricky Topic," can stand on its own. With our narrative now in hand, we can make an addition to any discussion about the meaning of IQ: IQ, along with GDP, is a metric of a population's economic ability. This association of IQ with economic success is consistent with the thesis developed in our narrative: high IQ and improved economic acumen coevolved hand in glove, and for this reason, IQ should be a good predictor of economic outcomes. Clearly, from figure 11.3, it is. IQ and GDP are joined at the hip, and Darwinian Theory provides the explanatory glue.

Many will not like this conclusion. Being partially deterministic, it is uncomfortable. There is another question, also with an uncomfortable conclusion. In a thousand years, what will this chart look like? Will we be more similar in economic performance and

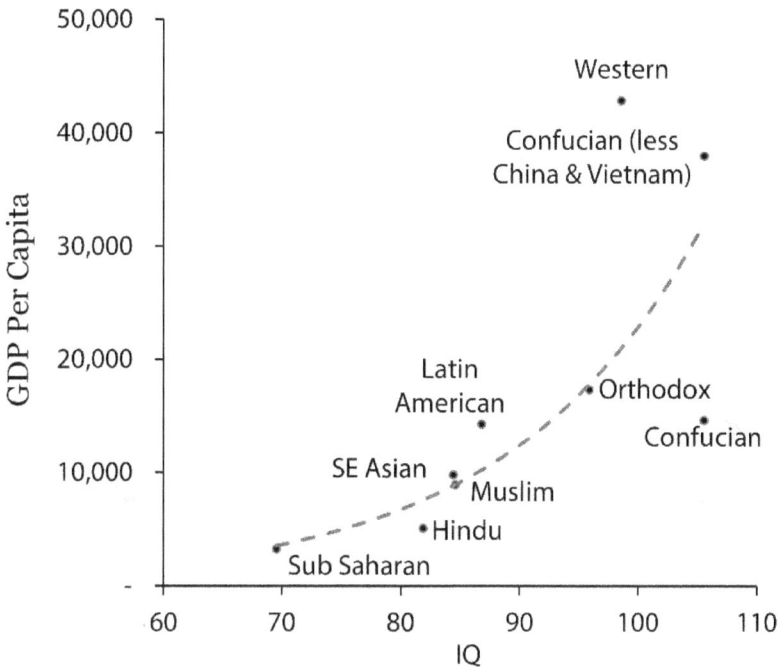

FIGURE 11.3   Wealth and the IQ of Nations (reproduced from chapter 11)

intelligence than now? Affecting the answer will be the fact that the Differencing Engine no longer works, an issue we will shortly discuss. We cannot depend on it to effect changes. We know where we are and, with our narrative, we have an idea of how and why we arrived at this point. How will we go forward? If we are to change, hopefully to become less different, there must be a reason. Because it is no longer, the reason will not be the Differencing Engine. If we are to change, there must be a reason, and to me, such a reason is not apparent. Perhaps to you, it is.

Even if there is a reason prodding our natures toward equality, the change will probably take considerable time. During our narrative, we made a couple of estimates for becoming civilized. For both the East and the West, that estimate was about a thousand years.

These conclusions of where we are—there is no change in sight, and if there is, change will take a very long time—are reality. Adults manage reality, making the best of the cards dealt.

There is one final point, a very important one needing illumination. We have been discussing populations, not individuals. Even though there are differences between populations, there will be individuals from a population on the left side of the graph just as prosperous and intelligent as others from the right side, just proportionally not as many.

Killing Biology

Introduction. The basis of evolution is differential survival and reproduction. Through natural selection, heritable traits promoting successful differential survival and reproduction increase in the population.

Before the Industrial Revolution, war, disease, and famine maintained a check on human population growth. Wealthy people, less influenced by these events, outsurvived and out-reproduced the poor, resulting in their heritable traits increasing in the population. The foregoing is the basis of our Differencing Engine.

"Growth of World Population," figure 3.3, makes obvious that, prior to the Industrial Revolution, our growth rate was close to zero. Disease, war, and famine kept us in check, and it was only after the advent of the Industrial Revolution that these natural constraints to growth lessened, allowing our population growth to explode. It did.

Darwinian fitness, at least as we have used it, no longer operates nearly as intensely as before. There is still war, but compared to our Neolithic days, death due to war is minimal. Famine only occurs in the poorest corners of the world, places where food aid cannot reach. Not only do we know what causes disease and how to prevent it but, once sick, we know how to effect cures. (However, do keep in mind that evolving bacteria and viruses could still defeat us.) With modern birth control methods, reproduction can be purely a matter of choice. We indulge in the procreative act but have control over its consequence. Another point is that we are now acquiring the ability to modify the code of life and engineer our own adaptations—no need to wait for evolution.

Before the Industrial Revolution, traits promoting survival and reproduction increased in the population. One consequence of the Industrial Revolution was the minimizing of constraints to growth. Currently, though constraints still exist, they are not nearly as formidable. After the Industrial Revolution, surviving and reproducing are becoming independent of heritable traits; now, they are more dependent on technology. Traits that used to increase in the population, because they increased fitness, no longer do. We have learned how to bypass Mother Nature; we have become masters of our own fitness. This is our new world, bright or not.

We developed our Differencing Engine with the premise that those with higher living standards had more surviving progeny. Before the Industrial Revolution, that was true. Today it is not. In the modern, advanced world, and in most of the remaining, everyone, poor or rich, survives. Additionally, based on personal choice, everyone can reproduce or not. It is now a matter of personal choice, not biological imperative. The key conclusion is that in our modern world, wealth does not confer fitness, at least not nearly to the extent it used to.

Importantly, our Differencing Engine no longer runs. Accordingly, populations will no longer change as they used to. Their engine of change has stopped.

Survival: War, famine, and disease. If we go back and revisit figure 5.1, "Percentage of Male Deaths due to Warfare," we will see that since our Neolithic days, death due to war has substantially decreased. In recent times, war does not appear to have significant consequence for survival and reproduction. Of course, that could quickly change given the technology available to modern warfare.

Do people still starve? Yes, but the problem is not food production. Globally, we produce enough and have the technology to meet our food needs. Consequently, when people starve, it is not for lack of food; it is the difficulty in getting food to them. However, in the long term, if the world population does not stop its increase, food availability could again be a problem.

Disease has been the biggest factor in limiting survival. For a time, cities had to import people from the rural environment just to maintain their population levels. This was due to death from disease limiting their population growth. Thankfully, that condition is behind us.

It is only somewhat recently that we have learned what causes disease and, with that, we now know how to prevent it. Both sanitation and vaccinations are prime examples. With continuing advancements in medicine, when we get sick, we often know how to cure. We still die, but when we do, it is well after we have had our go at reproducing. Any living past our reproductive period is more than *lagniappe*; in old age, we have use in raising and caring for our grandchildren. It is also a time to pass on what we know, provided the younger care to listen.

Have we achieved a disease-free utopia? No; some diseases, such as malaria, persist, but every year brings advancements, getting us closer to that goal.

We tend to put death on a personnel level; it is. However, our only real concern should be at the level of our species. As long as individuals reproduce, their later demise has no biological consequence. In our modern world, almost everyone gets the opportunity to pass his or her genes along. Before the Industrial Revolution, that was far from true.

Reproduction. We cannot forget about the other part of biological fitness, reproduction, and with modern science, we do not have to. We can indulge and succumb to all the biological imperatives, go through all the motions, but its result, the final choice, is now ours. Actually, with the availability of birth control pills, it is the choice of any woman living in a modern society; she can be in charge of the consequence of reproductive activity. Nevertheless, as abortion data show, not all choose to control their reproduction in this manner.

We do not have to reproduce. If we are selfish and choose not to aid our genes along their path into the future, we do not have to. The choice is ours. Why give up a hedonistic lifestyle or tranquil, laid-back Sunday mornings for some damn kids? The low fertility rate associated with advanced economies suggests that many simply opt out of this choice of life's rewards.

This is both interesting and troubling. By our study, the people of advanced, low-fertility economies are those possessing the heritable traits promoting wealth creation. Couple this with the observation that high fertility rates are associated with poor populations, those with a deficit of such traits, and we have a future with poor economic prospects. Additionally, there are far too many of us to consider the world returning to its preindustrial condition. As the ratio of wealth creators to non–wealth creators decreases, wealth available for redistribution to the poor will likewise decrease. Over the long term, and on a global scale, this does not portend well for our collective future.

Sexual selection. When we developed technology bypassing Mother Nature's rule, we did not eliminate all of natural selection, just the part relating to survival and reproduction. Sexual selection, a class of natural selection, still operates. As long as women believe that wealth is an indicator of fitness, through sexual selection, traits promoting wealth will increase.

Our Differencing Engine is in reverse. In the process of killing biology, we put the Differencing Engine into reverse at full throttle. The world is going backward. There was a time when increasing population levels were an indicator of Darwinian success. Nowadays, with the fertility levels of economically advanced nations decreasing and those at the other end of the economic spectrum increasing, that model no longer is true for us humans. On a global basis, the frequency of heritable traits promoting wealth creation is on the decrease, not increase, the opposite of when the Differencing Engine is in forward motion.

The survival and reproduction of Darwinian fitness are no longer properties associated with natural selection of heritable traits. They have become properties from our own cleverness and persistent desire to beat back reality. Mother Nature is still present, but her role is much smaller than before, as is the role of our Differencing Engine. Sometime between then and now, we broke biology. The Differencing Engine does not run, and traits promoting wealth creation are no longer a factor in natural selection; there is no need for them to be.

The cessation of the Differencing Engine is consequential; it is no longer available to improve the economic acumen and intelligence of populations. It will no longer cause change. It is like musical chairs: the music has ended and we are where we are, with chair or not. There is no more change toward traits promoting better economic performance or increased intelligence in view. The escalator has stopped. This is a truth economic developers working with poor nations must understand. People are different, and we are where we are. Any change in their nature will come only after a long time and, even then, only if there is reason.

There is an unstated consequence in the preceding discussion: we must live with who we are. If "who we are" includes differences between populations, and it does, we must understand what they are. This is not a place to be politically correct, and those of us who are adults should be able to handle truth's reality. What is important is that we manage our behavior accordingly and, in our management, consider our differences.

Who gets the girl now? In our beginning, when we were chimpanzee-like, the winner in the competition for sexual access to women was the alpha male. We suppressed his behavior, formed superorganismic societies, and access shifted to men performing altruistic acts for their community. In recent times, wealth was the key to competitive success. It increased fitness, and as women seek fit men as assurance their genes will pass on into the future, wealth became a sexual attractant. Now that wealth no longer increases fitness, will women still seek wealthy men? In the broader question, with no fitness differences, what will be the basis for mate selection?

This is more than a question from the ivory tower. In the past, much human behavior was a consequence of access to women. Our genetic code impels us to get our genes into the future, and, every now and then, the forces affecting our response change. We are in such a time. It is in our self-interest to understand what underlies human behavior, and we should work at understanding the implications of this change.

Changing behavior with change in wealth as a fitness indicator. Much of our narrative has been about wealth creation as a force in evolution. It made people fitter, resulting in evolutionary consequences. Now, we are all equally fit, and wealth creation no longer has its prior role. However, people still behave as if it does.

There are time lags to induced change. What we do not know is how long wealth creation will persist as a perceived fitness indicator. That is not the pertinent question, which is, what will be the economic consequences when wealth ceases its perceived association with fitness? Do hard work and diligence cease to be attributes driven by biology? We could generate a slew of similar questions, all related to this change, but we will not. We can save those for a later book.

Our self-directed future. We are not totally there yet, but, as just discussed, we essentially have mastery over our survival and reproduction. Our technology is making us all equally fit in the Darwinian sense. Are we yet gods? No, but within only the next millennium, we should be close, if not actually there.

That is not hubris. No one knows how much more there is to learn, but it must be finite. Are there things we can never know? Could there be unbridgeable gaps in our knowledge? Perhaps, but barring reality of the spiritual or truly mystical, it seems unlikely. The universe is not random; it is rational and determinable, and as humans, we will keep at it until nothing unknown remains. At that point, we will be gods, almost. We will still not be able to make a new universe or alter the rules of ours; only God can do that.

As suggested, the preceding assumption could be wrong, but, at a minimum, we will be able, within bounds, to engineer our genes and create whatever features we deem useful. We will be able to design babies with high IQ, charisma, and other desired traits. We are currently within a hundred years or so of that reality.

Taking the following steps to create life appears tractable. Someday we will probably be able to create life from basic ingredients, with the first model a bacteria-like, self-replicating pseudo-organism, just like in the beginning.

Will we ever be able to take the next step and create humans or even a simple organism? The answer to that question is no. Creation of even the simplest organism was via evolution, where, layer by layer and selected variation by selected variation, each organism became "created." To achieve that end, we would have to know both the nature of the changing environment and the specific variations selected in natural selection. Those are unknowable.

Making a copy of an existing organism is what might be possible. If we are able to make eggs and sperm, they can make life. Even if it turns out that we cannot, we will be able to modify what is to the extent that, in a practical sense, creating life becomes unimportant.

Some will not like the concept of humans creating life from basic ingredients, but that is the way life itself began. Unless there is a mystical aspect to creating self-replicating assemblages of molecules, we will. Admit it or not, many, if not most, hold mysterious the very essence of life. It is to them beyond the rational. To me, what is truly mysterious are all the rules in chemistry, quantum mechanics, physics, biology, probability theory, and so on, that concluded in me followed by my progeny. That is where mystery lives. Life is the consequence not of a creation but rather of a creator, One making all the rules, of which we are rule's consequence.

We are in new, uncharted waters. No longer do heritable traits such as hard work and intelligence have a role in our survival and reproduction. From a Darwinian sense, they do not increase our fitness. In the process of natural selection, they are no longer selected. Our fitness comes from the labs of pharmaceutical companies and the research conducted in the basements of important medical centers. It comes from organizations such as the United Nations, where preventing war is a priority. It comes from agronomists creating ever-higher-yielding varieties of basic foods. We are the agents of our fitness, the masters of our future.

The foregoing assumes that nature takes its course. As true masters of our universe, we can choose that course. Because we can, I suspect we will. The choice is ours. My only hope is that we do it with considered forethought. As the Chinese curse says, "may you live in interesting times." We are; most of us just do not know it yet.

Some will say we should not fool with things we do not understand. Following their dictate, we would not have solved the problem of famine or disease; after all, they are part of nature's design. Additionally, we would not have invented the birth control pill. We have already fooled Mother Nature, and besides, who would want to return to the not so good old days? We are not going to stop trying to improve the human condition, even if doing so might in actuality not be in our best interest.

Touching people. In chapter 1, the introduction to this book, there is a quotation from Jacob Bronowski; it is an emotional and impassioned plea for tolerance. Below is the section that follows, and on the TV program *The Ascent of Man,* Bronowski is standing in a field of ashes at Auschwitz. It is a fitting end to this section and our journey:

> It's said that science will dehumanize people and turn them into numbers. That's false, tragically false. Look for yourself. This is the concentration camp and crematorium at Auschwitz. This is where people were turned into numbers. Into this pond were flushed the ashes of some four million people. And that was not done by gas. It was done by arrogance, it was done by dogma, it was done by ignorance. When people believe that they have absolute knowledge, with no test in reality, this is how they behave. This is what men do when they aspire to the knowledge of gods.
>
> Science is a very human form of knowledge. We are always at the brink of the known; we always feel forward for what is to be hoped. Every judgment in science stands on the edge of error and is personal. Science is a tribute to what we can know although we are fallible. In the end, the words were said by Oliver Cromwell: "I beseech you in the bowels of Christ: Think it possible you may be mistaken."
>
> I owe it as a scientist to my friend Leo Szilard, I owe it as a human being to the many members of my family who died here, to stand here as a survivor and a witness. We have to cure ourselves of the itch for absolute knowledge and power. We have to close the distance between the push-button order and the human act. We have to touch people.[2]

Do not expect us humans to heed Bronowski's plea not to seek absolute knowledge and power. That would be *contra* our nature, and besides, using Bronowski's words, it is probably incurable. What we must do is recognize and acknowledge who we are and, with that knowledge, manage our journey into the future. Above all, however we choose to proceed in our self-directed future and all the while touching people along our way, we must shun arrogance, dogma, and ignorance.

Appendix: Data

Data used in this book are in the following table, which represents 145 nations. The abbreviations for Religions/Geographic Areas are as follows: CO = Confucian; HI = Hindu; IS = Islamic; LA = Latin American; OR = Orthodox Christian; SE = Southeast Asian; SS = sub-Saharan African; WE = Western.

The source of the data for GDP per capita (adjusted for natural resources) is the World Bank for 2013 (http://data.worldbank .org/). IQ data are from Richard Lynn and Tatu Vanhanen, *IQ and Global Inequality* (Augusta, Ga.: Washington Summit, 2006). Data for the Corruption Perception Index (2012) are from Transparency International (http://www.transparency.org/).

Country	Religion / Geography	GDP	IQ	CPI	Fert.	Urban
China	CO	11,635	105	40	1.56	54
Hong Kong	CO	53,367	108	75	1.12	100
Japan	CO	36,223	105	74	1.43	93
South Korea	CO	33,089	106	55	1.19	82
Singapore	CO	80,295	108	86	1.19	100
Taiwan	CO	43,678	105	61		78
Vietnam	CO	4,781	94	31	1.96	33
India	HI	5,158	82	36	2.47	32
Nepal	HI	2,135	78	31	2.29	18
Afghanistan	IS	1,923		8	5.05	26
Albania	IS	9,410		31	1.77	56
Algeria	IS	9,564		36	2.89	70

Country	Religion / Geography	GDP	IQ	CPI	Fert.	Urban
Azerbaijan	IS	10,903		28	2.00	54
Bahrain	IS	39,027		48	2.08	89
Bangladesh	IS	2,845		27	2.21	34
Bosnia and Herzegovina	IS	9,354		42	1.27	40
Egypt	IS	9,878	81	32	3.34	43
Iran	IS	9,736	84	25	1.73	73
Iraq	IS	8,970	87	16	4.61	69
Jordan	IS	11,559	84	45	3.47	83
Kazakhstan	IS	16,319		26	2.64	53
Kosovo	IS	8,774		33	2.16	
Kuwait	IS	33,548	86	43	2.15	98
Kyrgyz	IS	2,908		24	3.10	36
Lebanon	IS	17,173	82	28	1.70	88
Libya	IS	11,943		15	2.51	78
Morocco	IS	7,051	84	37	2.54	60
Oman	IS	26,755		47	2.82	77
Pakistan	IS	4,422	84	28	3.68	38
Qatar	IS	88,903	78	68	2.04	99
Saudi Arabia	IS	28,510		46	2.82	83
Tajikistan	IS	2,469		22	3.51	27
Tunisia	IS	10,435		41	2.20	67
Turkey	IS	18,695	90	50	2.09	73
Turkmenistan	IS	9,607		17	2.33	50
United Arab Emirates	IS	46,580		69	1.80	85
Uzbekistan	IS	4,129		17	2.30	36
Yemen	IS	3,334	85	18	4.28	34
Argentina	LA	22,098	93	34	2.34	92
Bolivia	LA	5,144	87	34	3.02	68
Brazil	LA	14,990	87	42	1.80	85
Chile	LA	18,407	90	71	1.77	89
Colombia	LA	11,246	84	36	1.92	76
Costa Rica	LA	13,733		53	1.84	76
Cuba	LA	20,788	85	46	1.62	77
Dominican Republic	LA	12,119	82	29	2.51	78

Country	Religion / Geography	GDP	IQ	CPI	Fert.	Urban
Ecuador	LA	9,039	88	35	2.57	64
El Salvador	LA	7,632		38	1.96	66
Guatemala	LA	7,005	79	29	3.26	51
Honduras	LA	4,129	81	26	2.44	54
Mexico	LA	15,125	88	34	2.27	79
Nicaragua	LA	4,332		28	2.30	58
Paraguay	LA	7,736	84	24	2.58	59
Peru	LA	10,631	85	38	2.48	78
Puerto Rico	LA	34,938	84	62	1.47	94
Uruguay	LA	19,442	96	73	2.03	95
Venezuela	LA	14,176	84	20	2.39	89
Belarus	OR	17,245		29	1.62	76
Bulgaria	OR	15,417	93	41	1.48	74
Cyprus	OR	23,564		63	1.46	67
Georgia	OR	7,095		49	1.82	53
Greece	OR	25,615	92	40	1.30	78
Macedonia	OR	11,859		44	1.51	57
Moldova	OR	4,678		35	1.26	45
Romania	OR	18,536	94	43	1.41	54
Russia	OR	20,527	97	28	1.70	74
Serbia	OR	12,642	89	42	1.43	55
Ukraine	OR	7,899		25	1.51	69
Cambodia	SE	2,906		20	2.68	21
Indonesia	SE	9,388	87	32	2.48	53
Lao	SE	4,075	89	26	3.06	38
Malaysia	SE	20,817	92	50	1.96	74
Philippines	SE	6,327	86	36	3.01	44
Sri Lanka	SE	9,679	79	37	2.11	18
Thailand	SE	13,731	91	35		
Timor-Leste	SE	2,162			5.20	32
Angola	SS	4,773		23	6.17	43
Benin	SS	1,687		36	4.85	44
Botswana	SS	16,207		64	2.86	57
Burkina Faso	SS	1,301		38	5.61	29
Burundi	SS	563		21	6.04	12
Cameroon	SS	2,578	64	25	4.78	54

Country	Religion / Geography	GDP	IQ	CPI	Fert.	Urban
Central African Republic	SS	509	64	25	4.37	40
Chad	SS	1,576		19	6.26	22
Congo, Dem. Rep.	SS	472	64	22	6.10	42
Congo, Rep.	SS	2,359	65	22	4.92	65
Cote d'Ivoire	SS	2,956		27	5.06	53
Gabon	SS	10,698		34	3.96	87
Gambia	SS	1,547		28	5.75	59
Ghana	SS	3,289	71	46	4.21	53
Guinea	SS	954	67	24	5.09	37
Guinea-Bissau	SS	1,130		19	4.91	49
Kenya	SS	2,704	72	27	4.41	25
Lesotho	SS	2,409		49	3.22	27
Liberia	SS	636		38	4.79	49
Madagascar	SS	1,275	82	28	4.47	34
Malawi	SS	667		37	5.22	16
Mali	SS	1,420		28	6.31	39
Mauritania	SS	1,708		30	4.66	59
Mauritius	SS	17,714	89	52	1.44	40
Mozambique	SS	946	64	30	5.42	32
Namibia	SS	9,400		48	3.56	46
Niger	SS	751		34	7.62	18
Nigeria	SS	4,740	69	25	5.71	47
Rwanda	SS	1,376		53	4.01	28
Senegal	SS	2,140		41	5.13	43
Sierra Leone	SS	1,746	64	30	4.75	40
South Africa	SS	11,992	72	42	2.39	64
South Sudan	SS	1,459		14	5.11	19
Sudan	SS	3,674	71	11	4.42	34
Swaziland	SS	6,054		39	3.33	21
Tanzania	SS	2,195	72	33	5.22	31
Togo	SS	1,267		29	4.66	39
Uganda	SS	1,457	73	26	5.87	16
Zambia	SS	3,164	71	38	5.43	40
Zimbabwe	SS	1,636	66	21	3.98	33
Australia	WE	40,098	98	81	1.86	89
Austria	WE	44,901	100	69	1.44	66

Country	Religion / Geography	GDP	IQ	CPI	Fert.	Urban
Belgium	WE	41,531	99	75	1.75	98
Canada	WE	40,795	99	81	1.61	82
Croatia	WE	20,988	90	48	1.46	59
Czech Republic	WE	28,902	98	48	1.46	73
Denmark	WE	43,038	98	91	1.67	88
Estonia	WE	25,281	99	68	1.52	68
Finland	WE	39,351	99	89	1.75	84
France	WE	37,557	98	71	1.99	79
Germany	WE	43,800	99	78	1.39	75
Hungary	WE	23,196	98	54	1.35	71
Ireland	WE	45,631	92	72	1.96	63
Italy	WE	35,005	102	43	1.39	69
Latvia	WE	21,925		53	1.52	67
Lithuania	WE	25,432	91	57	1.59	67
Netherlands	WE	45,700	100	83	1.68	90
New Zealand	WE	34,548	99	91	2.01	86
Norway	WE	58,485	100	86	1.78	80
Poland	WE	23,658	99	60	1.29	61
Portugal	WE	27,372	95	62	1.21	63
Slovak Republic	WE	26,364	96	47	1.34	54
Slovenia	WE	28,772	96	57	1.55	50
Spain	WE	33,059	98	59	1.27	79
Sweden	WE	44,155	99	89	1.89	86
Switzerland	WE	56,940	101	85	1.52	74
United Kingdom	WE	37,873	100	76	1.83	82
United States	WE	52,344	98	73	1.86	81

Notes

Preface

1. Edward O. Wilson, *Consilience: The Unity of Knowledge* (New York: Vantage Books, 1998), 8.
2. https://en.wiktionary.org/wiki/jack_of_all_trades,_master_of_none (accessed: 2016-05-29). Archived by WebCite® at http://www.webcitation .org/6hsKa5HrD.
3. Kyle Gibson, "Evolutionary Theory in Anthropology: Providing Ultimate Explanations for Human Behavior," http://digitalcommons.unl.edu/ nebanthro/64 (accessed: 2016-06-18). Archived by WebCite® at http://www .webcitation.org/6iMJCGOPs.

Chapter 1: Introduction

1. Jacob Bronowski, *The Ascent of Man* (New York: Little, Brown, 1973), 278.
2. See Ullica Segerstrale, "Defenders of the Truth: The Sociobiology Debate," May 31, 2001, for a detailed accounting. The term *tabula rasa* means "blank slate" in Latin. It is a theory that we are born without mental content. It posits that personality, behavior, and intelligence are formed after conception and have no genetic basis.
3. I suggest that Wilson's conclusions in the paragraph above are incorrect; political tests in the form of conformity to political correctness are a continuing problem. This continues to be true in both academia and government.
4. Edward O. Wilson, *Sociobiology: The New Synthesis,* 25th Anniversary ed. (Cambridge, Mass.: Belknap Press, 2000), vi.
5. David S. Landes, *The Wealth and Poverty of Nations: Why Some Are So Rich and Some So Poor* (New York: W. W. Norton, 1999), 516.
6. Ibid., 523.
7. David S. Landes, "Culture Makes Almost All the Difference," in *Culture Matters: How Values Shape Human Progress,* ed. Lawrence E. Harrison and Samuel P. Huntington (New York: Basic Books, 2000), 13.

8. Landes, *Wealth and Poverty*, 524.
9. Ibid., 175.
10. Lucien W. Pye, "'Asian Values': From Dynamos to Dominoes," in Harrison and Huntington, *Culture Matters*, 244–55.
11. David C. McClelland, *The Achieving Society* (New York: D. Van Nostrand, 2010).
12. Landes, *Wealth and Poverty*, 523.
13. Jared Diamond, *Guns, Germs, and Steel: The Fates of Human Societies* (New York: W. W. Norton, 1997), 25.
14. Ibid., 162–63.
15. Francis Fukuyama, *The Origins of Political Order: From Prehuman Times to the French Revolution* (New York: Farrar, Straus, and Giroux, 2011), 230.
16. L. H. Taylor, S. M. Latham, and M. E. Woolhouse, "Risk Factors for Human Disease Emergence," *Philosophical Transactions of the Royal Society, London, Series B* 356, no. 1411 (2001): 983–89.
17. Nicholas Wade, *A Troublesome Inheritance: Genes, Race and Human History* (New York: Penguin, 2014), 223.

Chapter 2: Laying the Foundation

1. Lawrence H. Keeley, *War before Civilization* (New York: Oxford University Press, 1996), 180.
2. The calculation is [((89% + 88% + 91% + 93%)/4) − ((65% + 73%)/2) = 21.25%].
3. Christopher Boehm, *Hierarchy in the Forest: The Evolution of Egalitarian Behavior* (Cambridge, Mass.: Harvard University Press, 1999).
4. Herbert Gintis, "Gene–Culture Coevolution and the Nature of Human Sociality," *Philosophical Transactions of the Royal Society, London, Series B* 366, no. 1566 (2011): 878–88.
5. Robert Brandon, "Natural Selection," in *The Stanford Encyclopedia of Philosophy*, Spring 2014 ed., ed. Edward Zalta, http://plato.stanford.edu/archives/spr2014/entries/natural-selection/.
6. Richard Lewontin, "The Genotype/Phenotype Distinction," in Zalta, *Stanford Encyclopedia of Philosophy*, Summer 2011 ed., http://plato.stanford.edu/archives/sum2011/entries/genotype-phenotype/.
7. Peter J. Richerson and Richard Boyd, "Evolution: The Darwinian Theory of Social Change—An Homage to Donald T. Campbell," in *Paradigms of Social Change: Modernization, Development, Transformation, Evolution*, ed. Waltraud Schelkle, Wolf-Hagen Krauth, Martin Kohli, and George Elwert (New York: St. Martin's Pres, 2000), 257–59.
8. Etienne Danchin and Richard H. Wagner, "Inclusive Heritability: Combining Genetic and Non-genetic Information to Study Animal Behavior and Culture," *Oikos* 119 (2010): 210–18.

9. Jonathon Pritchard and Anna Di Rienzo, "Adaptation—Not by Sweeps Alone," *Nature Reviews Genetics* 11, no. 10 (2010): 665–67.

Chapter 3: Inequalities' Consequences

1. Alexander Rosenberg and Frederic Bouchard, "Fitness," in Zalta, *Stanford Encyclopedia of Philosophy,* Fall 2015 ed., http://plato.stanford.edu/archives/fall2015/entries/fitness/.
2. Amy Chua, *World on Fire: How Exporting Free Market Democracy Breeds Ethnic Hatred and Global Instability* (New York: Random House, 2003), 3.
3. Roger Highfield, "Relative Wealth 'Makes You Happier,'" http://www.telegraph.co.uk/news/science/science-news/3315638/Relative-wealth-makes-you-happier.html (accessed: 2016–05–30). Archived by WebCite® at http://www.webcitation.org/6htr82Yn7.
4. Michael D. Tanner, "War on Poverty at 50," http://www.foxnews.com/opinion/2014/01/08/war-on-poverty-at-50-despite-trillions-spent-poverty-won.html (accessed: 2016–06–02). Archived by WebCite® at http://www.webcitation.org/6hyZm8dcm.
5. Michael Tanner and Charles Hughes, *The War on Poverty Turns 50: Are We Winning Yet?* (Cato Institute of Policy Analysis, October 20, 2014), no. 761.
6. William Easterly, *The White Man's Burden: Why the West's Efforts to Aid the Rest Have Done So Much Ill and So Little Good* (New York: Penguin Press, 2006), 4.
7. Tanner, "War on Poverty at 50."
8. Easterly, *White Man's Burden,* 2–5.
9. Ibid., 346–47.
10. https://en.wikipedia.org/wiki/Paradox_of_tolerance (accessed: 2016–07–28). Archived by WebCite® at http://www.webcitation.org/6jL8b11eW.
11. https://en.wikiquote.org/wiki/Karl_Popper (accessed: 2016–07–28). Archived by WebCite® at http://www.webcitation.org/6jL9B8mP4.

Chapter 4: Inequalities' Measure

1. Peter J. Richerson, Robert Boyd, and Robert L. Bettinger, "Was Agriculture Impossible during the Pleistocene but Mandatory during the Holocene? A Climate Change Hypothesis," *American Antiquity* 66, no. 3 (2001): 387–411.
2. Fukuyama, *Origins of Political Order,* 98.
3. "GDP per Capita, PPP (Current International $)," World Development Indicators database, World Bank, 2013.
4. GDP is the value of all goods and services produced in a country, and PPP is a correction that corrects for the fact that a dollar in Argentina goes a lot further than the same dollar in Norway. PPP is an estimate, not a "hard fact" like GDP.

5. Nathan Nunn and Nancy Qian, "The Columbian Exchange: A History of Disease, Food, and Ideas," *Journal of Economic Perspectives* 24, no. 2 (2010): 163–88.

6. George Gilder, *Wealth and Poverty* (Washington, D.C.: Regnery, 2012), 77–78.

7. Wealth of Religions is not the average of the GDP per person for the nations in a group. Instead, the GDP of all the nations of a bin were summed and then divided by the sum of the populations.

Chapter 5: Access to Women

1. Napoleon Chagnon, *Noble Savages: My Life among Two Dangerous Tribes— the Yanomamö and the Anthropologists* (New York: Simon and Schuster, 2013), 275–78.

2. Gregory Clark, *A Farewell to Alms: A Brief Economic History of the World* (New York: Prentice Hall, 1998), 112–23.

3. Chagnon, *Noble Savages,* 218.

4. Henry A. Kissinger, national security advisor and later secretary of state for the Nixon and Ford administrations, as quoted in the *New York Times,* October 28, 1973.

5. Napoleon Chagnon, *Yanomamö: The Fierce People* (New York: Holt, Reinhart, and Winston, 1968), 40.

6. Richard Dawkins, *The Selfish Gene* (New York: Oxford University Press, 1989).

7. Chagnon, *Noble Savages,* 394.

8. Richard Wrangham and Dale Peterson, *Demonic Males: Apes and the Origins of Human Violence* (New York: Mariner/Houghton Mifflin, 1996), 63.

9. Fukuyama, *Origins of Political Order,* 34.

10. Keeley, *War before Civilization,* 196–97. Chart derived from table 6.2, "Percentage of Deaths due to Warfare."

11. Bobbi S. Low, "An Evolutionary Perspective on War," in *Behavior, Culture, and Conflict in World Politics,* ed. W. Zimmerman and H. K. Jacobson (Ann Arbor: University of Michigan Press, 1993), 13–55.

12. Bruce Trigger, *Understanding Early Civilizations: A Comparative Study* (New York: Cambridge University Press, 2003), 629.

13. Ibid., 628.

14. Ibid.

15. Keeley, *War before Civilization,* 27–28.

16. Ibid., 186, table 2.2.

17. Chagnon, *Yanomamö,* 3.

18. Ibid., 123; Keeley, *War before Civilization,* 199.

19. Chagnon, *Yanomamö,* 123.

20. Keeley, *War before Civilization,* 68.

21. Chagnon, *Noble Savages,* 214–16.

22. Keeley, *War before Civilization,* 68.

23. Chagnon, *Noble Savages,* 277.

24. Ibid.

25. Low, "An Evolutionary Perspective on War."

26. Laura Betzig, "The Son Also Rises: Review of Geoffrey Clark's *A Farewell to Alms,*" *Evolutionary Psychology* 5 (2007): 733–39.

27. Clark, *A Farewell to Alms,* 7.

Chapter 6: Rediscovery of an Important Transition

1. M. C. Stiner, N. D. Munro, T. A. Surovell, E. Tchernov, and O. Bar-Yosef, "Paleolithic Population Growth Pulses Evidenced by Small Animal Exploitation," *Science, New Series* 283, no. 5399 (1999): 190–94; O. Bar-Yosef, "The Upper Paleolithic Revolution," *Annual Review of Anthropology* 31 (2002): 363–93.

2. Diamond, *Guns, Germs, Steel.*

3. R. M. Glasse, *Huli of Papua: A Cognate Descent System* (Paris: Moulton, 1968), 98.

4. Ibid., 107.

5. Ibid., 87.

6. Ibid., 48.

7. P. Sillitoe, "Big Men and War in New Guinea," *Man* 13, no. 2 (1978): 252–71.

8. Gregory Clark, "In Defense of the Malthusian Interpretation of History," *European Review of Economic History* 12 (2008): 175–99.

9. Chagnon, *Noble Savages,* 332.

10. V. Gordon Childe, *Man Makes Himself* (Nottingham, U.K.: Russell Press, 1936), 140–78. Childe develops the concepts for both the Neolithic Revolution and the Urban Revolution. In 1950, he papered a shorter and clearer précis, "The Urban Revolution," *Town Planning Review* 21 (1950): 3–17. This paper went on to become one of the most cited papers ever published by an archeologist.

11. Childe, "Urban Revolution."

Chapter 7: Group Selection and the Superorganism

1. B. Marilynn Brewer and Linnda R. Caporeal, "An Evolutionary Perspective on Social Identity: Revisiting Groups," in *Evolution and Social Psychology,* ed. Mark Schaller, Jeffry A. Simpson, and Douglas T. Kenrick (New York: Psychology Press, 2014), 143–61.

2. Christopher Boehm, "Impact of the Human Egalitarian Syndrome on Darwinian Selection Mechanics," *American Naturalist* 150, no. S1 (July 1997): S100–S121; Edward O. Wilson, *The Social Conquest of Earth* (New York: Liveright, 2012), 170–88; Elliott Sober and David S. Wilson, *Unto Others: The Evolution and Psychology of Unselfish Behavior* (Cambridge, Mass.: Harvard University Press, 1998), 329–37; Peter J. Richerson, "Comment on Steven

Pinker's *Edge* essay The False Allure of Group Selection," http://www.des
.ucdavis.edu/faculty/Richerson/Comment%20on%20Pinker%20Edge%20essay
.pdf (accessed: 2016-07-28), archived by WebCite® at http://www.webcitation
.org/6jLe8j47H; Robert Boyd and Peter J. Richerson, "Transmission Coupling
Mechanisms: Cultural Group Selection," *Philosophical Transactions of the Royal
Society, London, Series B* 365, no. 1559 (2010): 3787–95; M. A. Nowak, C. E.
Tarnita, and E. O. Wilson, "The Evolution of Eusociality," *Nature* 466, no. 7310
(2010): 1057–62; Jonathan Haidt, *The Happiness Hypothesis: Finding Modern
Truth in Ancient Wisdom* (New York: Basic Books, 2006), 230–35; Brewer and
Caporeal, "An Evolutionary Perspective on Social Identity," 143–61.
3. R. A. Fisher, *The Genetical Theory of Natural Selection* (Oxford: Clarendon
Press, 1930), 159; J. B. S. Haldane, *The Causes of Evolution* (London:
Longmans, Green, 1932); Haldane, "Population Genetics," *New Biology* 18
(1955): 34–51.
4. W. D. Hamilton, "The Genetical Evolution of Social Behaviour I & II,"
Journal of Theoretical Biology 7, no. 1 (1964): 1–52.
5. Dawkins, *Selfish Gene*, xxi.
6. Wilson, *Social Conquest of Earth*, 170–88.
7. A review of Wilson's book by Herbert Gintis having the appropriate title
Clash of the Titans provides a good overview of the debate. Gintis, "Clash of
the Titans," *Bioscience* 62, no. 11 (2012): 987–91.
8. Sober, *Unto Others*, 57.
9. Bronisław Malinowski, *Argonauts of the Western Pacific: An Account of
Native Enterprise and Adventure in the Archipelagos of Melanesian New Guinea*
(New York: E. P. Dutton, 1932). A copy of Malinowski's opus is located at
https://openlibrary.org/books/OL7149116M/Argonauts_of_the_western_
Pacific.
10. Ibid., 61–62.
11. Selin Kesebir, "The Superorganism Account of Human Sociality: How and
When Human Groups Are Like Beehives," *Personality and Social Psychology
Reviews* 16, no. 3 (2012): 233–61.
12. Karl Polanyi, *The Great Transformation: The Economic and Political Origins
of Our Time* (New York: Beacon Press, 2001), 276–80.
13. Adam Smith, *The Theory of Moral Sentiments,* part 1, chapter 1 (1759).
14. Thomas Hobbes, *Leviathan,* part 1, chapter 13 (1651).
15. Hobbes had it wrong in thinking that pre-civilized man was solitary. He
was anything but, as the individual was part of a colonial superorganism. In a
Hobbesian sense, the collective, survival machine was solitary. It was only after
the founding of the state that the individual existed alone in Hobbes's "state of
nature."
16. Boehm, *Hierarchy in the Forest,* 194–95.
17. Keeley, *War before Civilization,* 4.

18. Edward O. Wilson, "Wilson Says Greedy People, Altruistic Groups Win," http://www.bloomberg.com/news/articles/2012-04-17/wilson-says-greedy-people-altruistic-groups-win (accessed: 2016-07-31). Archived by WebCite® at http://www.webcitation.org/6jPu32rIc.

19. Boehm, *Hierarchy in the Forest,* 61.

20. Ibid., 10.

21. Ibid., 61.

22. Ibid., 84.

23. Harold Schneider, *Livestock and Equality in East Africa: The Economic Basis for Social Structure* (Bloomington: Indiana University Press, 1979), as cited in Boehm, *Hierarchy in the Forest,* 105.

24. Boehm, *Hierarchy in the Forest,* 87.

25. Ibid.

26. Wilson, "Wilson Says."

27. Keeley, *War before Civilization,* 37; Steven Pinker, *The Better Angles of Our Nature: Why Violence Has Declined* (New York: Penguin Books, 2011), 16–47.

28. Keeley, *War before Civilization,* 196–97, table 6.2; Pinker, *Better Angles of Our Nature,* 692–96.

29. David M. Buss, "Sexual Selection," https://www.edge.org/response-detail/11226 (accessed: 2016-07-31). Archived by WebCite® at http://www.webcitation.org/6jPutuhep.

30. Menelaos Apostolou, *Sexual Selection under Parental Choice: The Evolution of Human Mating Behavior* (New York: Psychology Press, 2014), 40.

31. Wilson, *Social Conquest of Earth,* 72.

Chapter 8: Nature of the Superorganism

1. Kesebir, "Superorganism Account of Human Sociality."

2. Ibid.

3. Ibid.

4. Wilson, "Wilson Says."

5. Boehm, *Hierarchy in the Forest,* 235.

6. Charlie L. Hardy and Mark Van Vugt, "Nice Guys Finish First: The Competitive Altruism Hypothesis," *Personality and Social Psychology Bulletin* 32, no. 10 (2006): 1402–13.

7. https://en.wikipedia.org/wiki/Cheating_(biology) (accessed: 2016-07-31). Archived by WebCite® at http://www.webcitation.org/6jPvM5VQd.

8. Chris Baker, *Cultural Studies: Theory and Practice* (London: Sage, 2005), 448.

9. Malinowski, *Argonauts of the Western Pacific,* 22.

10. Ibid., 326.

11. Ibid., 62.

12. Ibid., 258.

13. Wade, *Troublesome Inheritance,* 128.

14. Jared Diamond, *The World until Yesterday: What We Can Learn from Traditional Societies* (New York: Penguin Group, 2012), 91–92.

Chapter 9: Consequence of the Superorganism

1. http://www.unitedhoumanation.org/about (accessed: 2016–08–01). Archived by WebCite® at http://www.webcitation.org/6jREfvOFo.
2. https://www.cia.gov/library/publications/the-world-factbook/geos/ni.html (accessed: 2016–08–01). Archived by WebCite® at http://www.webcitation.org/6jRGQF3iF.
3. Daniel G. Bates and Fred Plog, *Cultural Anthropology*, 3rd ed. (New York: McGraw-Hill, 1976), 6.
4. Robert Boyd and Peter J. Richerson, *The Origin and Evolution of Cultures* (New York: Oxford University Press, 2005), 6.
5. Ibid., 204.
6. Ibid., 264.
7. Polanyi, *The Great Transformation*, 48.
8. Ibid., 52.
9. https://en.wikipedia.org/wiki/Economic_problem (accessed: 2016–07–31). Archived by WebCite® at http://www.webcitation.org/6jQGfQjbk.
10. Malinowski, *Argonauts of the Western Pacific*, 60.
11. Leopold Pospisil, *Kapauku Papuan Economy* (Toronto: Burns and MacEachern, 1963), 145.
12. Robert L. Carneiro, "Slash-and-Burn Cultivation among the Kuikuru and Its Implications for Cultural Development in the Amazon Basin," in *Native South Americans*, ed. Patricia J. Lyon (Boston: Little, Brown, 1961), 122–32.
13. Marshall Sahlins, *Stone Age Economics* (New York: Aldine de Gruyter, 1972), 1–39.
14. Bronisław Malinowski, "The Primitive Economics of the Trobriand Islanders," *Economic Journal* 31 (1921): 1–16.
15. Polanyi, *The Great Transformation*, 171–72.
16. Ronald E. Seavoy, *Famine in Peasant Societies* (New York: Greenwood Press, 1986), 219.
17. Ibid., 207.

Chapter 10: Consequences of the Urban Revolution

1. Helmut K. Anheier and Mark Juergensmeyer, *Encyclopedia of Global Studies* (London: Sage, 2012), 1721.
2. Clark, *A Farewell to Alms*, 7.
3. Daniel Nettle and Thomas V. Pollet, "Natural Selection on Male Wealth in Humans," *American Naturalist* 172, no. 5 (2008).
4. Brandon, "Natural Selection."
5. Ibid.
6. Ibid.

7. Ibid.

8. Daron Acemoglu and James A. Robinson, *Why Nations Fail: The Origins of Power, Prosperity, and Poverty* (New York: Crown Business, 2002), 74.

9. Ibid., 76.

10. Peter J. Richerson and Robert Boyd, "Evolution: The Darwinian Theory of Social Change, an Homage to Donald T. Campbell," in *Paradigms of Social Change: Modernization, Development, Transformation, Evolution,* ed. W. Schelkle, W. H. Krauth, M. Kohli, and G. Ewarts (Frankfurt: Campus, 2000), 257–82.

11. Pliny the Elder, *Natural History* 12.84.

Chapter 11: The Differencing Engine and the Peacock's Tail

1. N. J. Macintosh, *IQ and Human Intelligence,* 2nd ed. (New York: Oxford University Press, 2011), 56–57.

2. Richard Lynn and Tatu Vanhanen, *IQ and Global Inequality* (Augusta, Ga.: Washington Summit, 2006), 296–310.

3. There is one outlier in figure 11.2: Qatar. It is the data point at IQ = 78, GDP = 88,000. This suggests that oil wealth can bring a lot more to GDP than just "black gold." China is the point at IQ = 105, GDP = 11,635.

4. https://en.wikipedia.org/wiki/Correlation_does_not_imply_causation (accessed: 2016-07-31). Archived by WebCite® at http://www.webcitation. org/6jQ50cW6N.

5. Gregory Cochrane and Henry Harpending, *The 10,000 Year Explosion: How Civilization Accelerated Human Evolution* (New York: Basic Books, 2009), 188.

6. Ibid., 194.

7. Ibid., 197.

8. Ibid., 199–200.

9. Akbar Ganji, "Revolutionary Pragmatists: Why Iran's Military Won't Spoil Détente with the U.S.," https://www.foreignaffairs.com/articles/middle-east/2013-11-10/revolutionary-pragmatists (accessed: 2016-07-31). Archived by WebCite® at http://www.webcitation.org/6jQ4lXT1d.

10. Acemoglu and Robinson, *Why Nations Fail,* 73.

11. Mark Van Vugt and Anjana Ahuja, *Why Some People Lead, Why Others Follow, and Why It Matters* (Toronto: Random House Canada, 2011), 123.

12. Acemoglu and Robinson, *Why Nations Fail,* 430.

13. Michele Boldrin, David K. Levine, and Salvatore Modica, "A Review of Acemoglu and Robinson's Why Nations Fail," http://www.dklevine.com/general/aandrreview.pdf (accessed: 2016-07-31). Archived by WebCite® at http://www.webcitation.org/6jQ1BEHwU.

14. http://www.transparency.org/cpi2012/results#myAnchor1 (accessed: 2016-07-31). Archived by WebCite® at http://www.webcitation.org/6jQ0w2z10.

15. Wilson, *Social Conquest of Earth,* 170–88.

16. Acemoglu and Robinson, *Why Nations Fail,* 427.

Chapter 12: Applying the Model

1. This is significant from a global perspective, for if China increases its GDP per capita to that of the United States, the world economy will have increased by greater than 80 percent.

2. Chua, *World on Fire*, 1.

3. https://en.wikipedia.org/wiki/Confucius (accessed: 2016-07-31). Archived by WebCite® at http://www.webcitation.org/6jPzvi7O6.

4. http://confucius-1.com/teachings/ (accessed: 2016-07-31). Archived by WebCite® at http://www.webcitation.org/6jQ0Ct3Pr.

5. Ibid.

6. Jeffrey Riegel, "Confucius," in Zalta, *Stanford Encyclopedia of Philosophy*, Summer 2013 ed., http://plato.stanford.edu/archives/sum2013/entries/confucius/.

7. Ibid.

8. Ibid.

9. Fukuyama, *Origins of Political Order*, 98.

10. Ibid.

11. Ibid., 56.

12. Ibid., 51.

13. Fareed Zakaria, "Culture Is Destiny: A Conversation with Lee Kuan Yew," *Foreign Affairs*, March/April 1994, 109–26.

14. Samuel Brittan, "Essays, Moral, Political and Economic," *Hume Papers on Public Policy* 6, no. 4 (1998): 4.

15. Fukuyama, *Origins of Political Order*, 118.

16. Ibid., 103.

17. Ibid., 101.

18. Ibid., 295.

19. Acemoglu and Robinson, *Why Nations Fail*, 445–46.

20. Romila Thapar, *Early India: From the Origins to AD 1300* (Berkeley: University of California Press, 2001), 62–68.

21. Fukuyama, *Origins of Political Order*, 152.

22. Wade, *Troublesome Inheritance*, 141.

23. Nicholas Tarling, ed., *The Cambridge History of Southeast Asia*, vol. 1, *From Early Times to c. 1500* (Cambridge: Cambridge University Press, 1991), 185.

24. Ibid., 1:185.

25. Ibid., 1:192.

26. Ibid., 1:194.

27. Ibid.

28. Ibid., 1:158.

29. Ibid., 1:200.

30. Ibid., 1:208.

31. Ibid., 1:215.

32. Ibid., 1:226.

33. Damian Evans, Christophe Pottier, Roland Fletcher, Scott Hensley, Ian Tapley, Anthony Milne, and Michael Barbetti, "A Comprehensive Archaeological Map of the World's Largest Preindustrial Settlement Complex at Angkor, Cambodia," *Proceedings of the National Academy of Sciences* 104, no. 36 (2007): 14277–82.

34. Tarling, *Cambridge History of Southeast Asia,* 1:158.

35. Ibid., 1:160.

36. Ibid., 1:163.

37. Evans et al., "A Comprehensive Archaeological Map."

38. Tarling, *Cambridge History of Southeast Asia,* 1:163.

39. Ibid.

40. Ibid., 1:167.

41. Nicholas Tarling, ed., *The Cambridge History of Southeast Asia,* vol. 2, *From c. 1500 to c. 1800* (Cambridge: Cambridge University Press, 1992), 59.

42. Ibid., 2:73.

43. Ibid., 2:61.

44. Ibid.

45. Ibid., 2:73.

46. Ibid., 2:117.

47. Ibid.

48. Ibid., 2:96, 101.

49. Ibid., 2:117–18.

50. Ibid., 2:27.

51. Ibid., 2:63.

52. Fukuyama, *Origins of Political Order,* xi–xiii.

53. https://en.wikipedia.org/wiki/Mamluk (accessed: 2016–07–31). Archived by WebCite® at http://www.webcitation.org/6jPiRDIk3.

54. Fukuyama, *Origins of Political Order,* 192.

55. Chris Wickham, *The Inheritance of Rome: Illuminating the Dark Ages 400–1000* (New York: Penguin Books, 2009), 24.

56. Ibid.

57. Ibid., 34.

58. https://en.wikipedia.org/wiki/History_of_Rome (accessed: 2016–07–31). Archived by WebCite® at http://www.webcitation.org/6jPiBcdfv.

59. Fukuyama, *Origins of Political Order,* 21.

60. Ibid., 238.

61. Ibid., 238–39.

62. Ibid., 231.

63. Ibid., 374.

64. Marc Bloch, *Feudal Society* (Chicago: Chicago University Press, 1968), 148, as quoted in Fukuyama, *Origins of Political Order,* 236.

65. Fukuyama, *Origins of Political Order,* 236.

66. Henri Pirenne, *Medieval Cities: Their Origins and Revival of Trade,* trans. Frank D. Halsey (Princeton, N.J.: Princeton University Press, 1925), 41.

67. Ibid., 43.

68. Fukuyama, *Origins of Political Order,* 373.

69. Jerome Blum, *The European Peasantry from the Fifteenth to the Nineteenth Century* (Washington, D.C.: Service Center for Teachers of History, 1960), 15–16, as quoted in Fukuyama, *Origins of Political Order,* 376.

70. Pirenne, *Medieval Cities,* 95.

71. Adam Smith, *The Wealth of Nations,* Books I–III, edited and with an introduction and notes by Andrew Skinner (New York: Penguin Books, 1999), 499–500.

72. Pirenne, *Medieval Cities,* 122.

73. Landes, *Wealth and Poverty,* 36.

74. Ibid., 37.

75. Pirenne, *Medieval Cities,* 125.

76. Fukuyama, *Origins of Political Order,* 410.

77. Thomas Sowell, *Migration and Cultures: A World View* (New York: Basic Books, 1996), 54.

78. Fukuyama, *Origins of Political Order,* 387.

79. J. L. I. Fennell, *Ivan the Great of Moscow* (London: St. Martin's Press, 1961), 354, as quoted in https://en.wikipedia.org/wiki/Ivan_III_of_Russia (accessed: 2016-07-30). Archived by WebCite˙ at http://www.webcitation.org/6jOcfXlJk.

80. Fukuyama, *Origins of Political Order,* 332.

81. Ibid., 331.

82. Ibid.

83. Ibid., 375.

84. Ibid.

85. Ibid.

86. Ibid., 241.

87. Ibid., 412.

88. Ibid., 413.

89. Ibid., 392.

90. Ibid., 377.

91. Ibid., 412–13.

92. Thomas Sowell, *Conquests and Cultures: An International History* (New York: Basic Books, 1998), 209.

93. Ibid.

94. Ibid.

95. Ibid.

96. Jerome Blum, *Lord and Peasant in Russia, from the Ninth to the Nineteenth Century* (Princeton, N.J.: Princeton University Press, 1961), 370, as quoted in Fukuyama, *Origins of Political Order,* 376.

97. Sowell, *Conquests and Cultures,* 207.

98. Daniel Thorner, Basil Kerblay, and R. E. F. Smith, eds., *A. V. Chayanov on the Theory of Peasant Economy* (Madison: University of Wisconsin Press, 1968), lxv.

99. Ibid., xv.

100. Peter Wehrheim and Peter Wobst, "The Economic Role of Russia's Subsistence Agriculture in the Transition Process," *Agricultural Economics* 33, no. 1 (2005): 91–105.

101. Vladimir Bogdanovskii, "Rural and Agricultural Markets," in *Russia's Agriculture in Transition: Factor Markets and Constraints on Growth*, ed. Zvi Lerman, Rural Economies in Transition (Lanham, Md.: Lexington Books/ Rowman and Littlefield, 2007), 240.

102. Lilia Ovcharova, "Russia's Middle Class: At the Centre or on the Periphery of Russian Politics?," http://www.iss.europa.eu/publications/detail/article/ russias-middle-class-at-the-centre-or-on-the-periphery-of-russian-politics/ (accessed: 2016–07–30). Archived by WebCite® at http://www.webcitation .org/6jOHZPV6v.

103. Sowell, *Conquests and Cultures*, 252.

104. Ibid., 250.

105. Ibid., 249.

106. Ibid., 250.

107. Ibid., 251.

108. Ibid., 249.

109. Ibid., 251.

110. Ibid., 254.

111. Ibid., 255.

112. Ibid.

113. Ibid.

114. Ibid.

115. Nathan Nunn and Nancy Qian, "The Columbian Exchange: A History of Disease, Food, and Ideas," *Journal of Economic Perspectives* 24, no. 2 (2010): 163–88.

116. Landes, *Wealth and Poverty*, 312.

117. Ibid.

118. Sowell, *Conquests and Cultures*, 299.

119. Landes, *Wealth and Poverty*, 312.

120. Robert A. Maryks, "Purity of Blood," http://www.oxfordbibliographies .com/view/document/obo-9780195399301/obo-9780195399301-0101. xml (accessed 2016–07–29), doi:10.1093/obo/9780195399301–0101; https://en.wikipedia.org/wiki/Casta (accessed: 2016–07–30). Archived by WebCite® at http://www.webcitation.org/6jOGFXXzQ.

121. https://en.wikipedia.org/wiki/Casta (accessed: 2016–07–30). Archived by WebCite® at http://www.webcitation.org/6jOGFXXzQ.

122. Landes, *Wealth and Poverty*, 311.

123. Sowell, *Conquests and Cultures*, 297.

124. https://en.wikipedia.org/wiki/Encomienda (accessed: 2016–07–30). Archived by WebCite˚ at http://www.webcitation.org/6jOFgCbqk.

125. Sowell, *Conquests and Cultures*, 297.

126. Fukuyama, *Origins of Political Order*, 439.

127. Landes, *Wealth and Poverty*, 495.

128. Jeffry Herbst, *States and Power in Africa* (Princeton, N.J.: Princeton University Press, 2000), xxvii.

129. Ivor Wilks, "On Mentally Mapping Greater Asante: A Study of Time and Motion," *Journal of African History* 33 (1992): 175–90.

130. Herbst, *States and Power in Africa*, 49.

131. Ibid., 11.

132. Ibid., 39.

133. Ibid., 20.

134. Ibid., 52.

135. Ibid., 49.

136. Wade, *Troublesome Inheritance*, 135.

137. Robert M. Maxon, "Social and Cultural Changes," in *Decolonization and Independence in Kenya (1940–1993)*, ed. B. A. Ogot and W. R. Ochieng (Athens: Ohio University Press, 1995), 116.

138. Ibid., 124.

139. Ibid., 123.

140. Ibid.

141. https://www.cia.gov/library/publications/the-world-factbook/rankorder/2127rank.html (accessed: 2016–07–30). Archived by WebCite˚ at http://www.webcitation.org/6jOEjvQOF.

142. Maxon, "Social and Cultural Changes," 123.

143. Ibid.

144. Ken Post, "Peasantization in Western Africa," in *African Social Studies: A Radical Reader*, ed. Peter C. W. Gutkind and Peter Waterman (New York: Monthly Review Press, 1977), 241.

145. Monica Das Gupta, "Kinship Systems and Demographic Regimes," in *Anthropological Demography: Toward a New Synthesis*, ed. David I. Kertzer and Tom Fricke (Chicago: University of Chicago Press, 1997), 40.

146. http://www.cnpp.usda.gov/tools/CRC_Calculator/ (accessed: 2016–07–30). Archived by WebCite˚ at http://www.webcitation.org/6jOCojgR1.

147. April A. Gordon, *Transforming Capitalism and Patriarchy: Gender and Development in Africa* (Boulder, Colo.: Lynne Rienner, 1996), 47–48.

Chapter 13: Lagniappe

1. Wade, *Troublesome Inheritance*, 237.

2. Bronowski, *Ascent of Man*, 278.

www.ingramcontent.com/pod-product-compliance
Lightning Source LLC
Chambersburg PA
CBHW060959280326
41935CB00009B/760